Lecture Notes of the Institute for Computer Sciences, Social Informatics and Telecommunications Engineering 148

More information about this series at http://www.springer.com/series/8197

Igor Bisio (Ed.)

Personal Satellite Services

Next-Generation Satellite Networking and Communication Systems

6th International Conference, PSATS 2014
Genova, Italy, July 28–29, 2014
Revised Selected Papers

 Springer

Editor
Igor Bisio
Department of Telecommunication,
  Electronic, Electrical Engineering
  and Naval Architecture
University of Genova
Genova
Italy

ISSN 1867-8211   ISSN 1867-822X (electronic)
Lecture Notes of the Institute for Computer Sciences, Social Informatics
and Telecommunications Engineering
ISBN 978-3-319-47080-1   ISBN 978-3-319-47081-8   (eBook)
DOI 10.1007/978-3-319-47081-8

Library of Congress Control Number: 2016953295

Printed on acid-free paper

This Springer imprint is published by Springer Nature
The registered company is Springer International Publishing AG
The registered company address is: Gewerbestrasse 11, 6330 Cham, Switzerland

Message from the General Chairs

It is our great pleasure to welcome you to the proceedings of the 6th International Conference on Personal Satellite Services (PSATS), held in Genoa, Italy. PSATS represents one of the most interesting gatherings of researchers and industry professionals in the field of satellite and space communications, networking, and services in the world. The sixth edition of the PSATS conference was no exception and brought together delegates from around the globe to discuss the latest advances in this vibrant and constantly evolving field.

The program included interesting keynote speeches from a highly innovative start-up, Outernet, presented by its founder Sayed Karim; from a big enterprise in the field, Ansaldo STS S.p.A., presented by Senior Vice-President Francesco Rispoli; and from academia, presented by two experts in the field of satellite and space networking, Prof. Franco Davoli and Prof. Mario Marchese, both from the University of Genoa. Ansaldo STS S.p.A. and the University of Genoa, together with the EIA, sponsored the conference and the success of the event is due in great part to their contributions.

The delegates of PSATS 2014 discussed and presented the latest advances in next-generation satellite networking and communication systems. A diverse range of topics from nano-satellites, satellite UAVs, as well as protocols and applications were featured at the conference. However, the major transformation is likely to be due to the increased capability of satellite technologies and their infiltration in new application domains with a profound impact on many sectors of our economy and the potential to lead to new paradigms in services and transportation. These were the messages derived from the presentation of the ten high-quality accepted papers, which represent approximately 50 % of the submitted works.

Finally, the program also included two very exciting demos. The first, introduced by Prof. Carlo Caini, from the University of Bologna, was about delay-tolerant networks; the second, prepared by the Digital Signal Processing Laboratory of the University of Genoa (www.dsp.diten.unige.it), was on application layer coding for video streaming with mobile terminals over satellite/terrestrial networks.

In addition to the stimulating program of the conference, the delegates enjoined Genoa and the Ligurian Riviera, with its tourist attractions, the diversity and quality of its cuisine, and world-class facilities. It is an unforgettable place to visit. It was a pleasure, therefore, to bring the conference attendants to Genoa and its surroundings to enjoy the vibrant atmosphere of the city.

Finally, it was a great privilege for us to serve as the general chairs of PSATS 2014 and it is our hope that you find the conference proceedings stimulating.

July 2014

Igor Bisio
Nei Kato

Organization

General Chairs

Igor Bisio	University of Genoa, Italy
Nei Kato	Tohoku University, Japan

TPC Chairs

Tomaso de Cola	German Aerospace Center, Germany
Song Guo	The University of Aizu, Japan

Industrial Chairs

Francesco Rispoli	Ansaldo STS, Italy
Chonggang Wang	InterDigital, USA

Publicity Chairs

Ruhai Wang	Lamar University, USA
Mauro De Sanctis	University of Rome Tor Vergata, Italy

Demos and Tutorial Chairs

Scott Burleigh	NASA Jet Propulsion Laboratory, USA
Carlo Caini	University of Bologna, Italy

Publications Chair

Giuseppe Araniti	University Mediterranea of Reggio Calabria, Italy

Local Organizing Chair

Marco Cello	University of Genoa, Italy

Website Chairs

Stefano Delucchi	University of Genoa, Italy
Andrea Sciarrone	University of Genoa, Italy

Steering Committee

Imrich Chlamtac   Create-Net, Italy (Chair)
Kandeepan   RMIT, Australia
  Sithamparanathan
Agnelli Stefano   ESOA/Eutelsat, France
Mario Marchese   University of Genoa, Italy

Advisory Committee

Giovanni Giambene   University of Siena, Italy
Fun Hu   University of Bradford, UK
Vinod Kumar   Alcatel-Lucent, France

Contents

Satellite Networking in the Context of Green, Flexible and Programmable
Networks.. 1
  Franco Davoli

Extended Future Internet: An IP Pervasive Network Including
Interplanetary Communication?................................ 12
  Mario Marchese

A Fast Vision-Based Localization Algorithm for Spacecraft in Deep Space.... 22
  Qingzhong Liang, Guangjun Wang, Hui Li, Deze Zeng, Yuanyuan Fan,
  and Chao Liu

Performance Evaluation of HTTP and SPDY Over a DVB-RCS Satellite
Link with Different BoD Schemes.............................. 34
  Luca Caviglione, Alberto Gotta, A. Abdel Salam, Michele Luglio,
  Cesare Roseti, and F. Zampognaro

Telecommunication System for Spacecraft Deorbiting Devices 45
  Luca Simone Ronga, Simone Morosi, Alessio Fanfani,
  and Enrico Del Re

Quality of Service and Message Aggregation in Delay-Tolerant
Sensor Internetworks.. 58
  Edward J. Birrane III

Virtualbricks for DTN Satellite Communications Research and Education ... 76
  Pietrofrancesco Apollonio, Carlo Caini, Marco Giusti,
  and Daniele Lacamera

Research Challenges in Nanosatellite-DTN Networks 89
  Marco Cello, Mario Marchese, and Fabio Patrone

A Dynamic Trajectory Control Algorithm for Improving the Probability
of End-to-End Link Connection in Unmanned Aerial Vehicle Networks..... 94
  Daisuke Takaishi, Hiroki Nishiyama, Nei Kato, and Ryu Miura

Hybrid Satellite-Aerial-Terrestrial Networks for Public Safety............ 106
  Ying Wang, Chong Yin, and Ruijin Sun

Satellites, UAVs, Vehicles and Sensors for an Integrated Delay Tolerant
Ad Hoc Network 114
  Manlio Bacco, Luca Caviglione, and Alberto Gotta

Smartphones *Apps* Implementing a Heuristic Joint Coding
for Video Transmissions Over Mobile Networks..................... 123
   Igor Bisio, Fabio Lavagetto, Giulio Luzzati, and Mario Marchese

Author Index 133

Satellite Networking in the Context of Green, Flexible and Programmable Networks

(Invited Paper)

Franco Davoli[✉]

Department of Electrical, Electronic, Telecommunications Engineering
and Naval Architecture (DITEN), University of Genoa/CNIT – University
of Genoa Research Unit, Via Opera Pia 13, 16145 Genoa, Italy
franco.davoli@unige.it

Abstract. In order to support heterogeneous services, using the information generated by a huge number of communicating devices, the Future Internet should be more energy-efficient, scalable and flexible than today's networking platforms, and it should allow a tighter integration among heterogeneous network segments (fixed, cellular wireless, and satellite). Flexibility and in-network programmability brought forth by Software Defined Networking (SDN) and Network Functions Virtualization (NFV) appear to be promising tools for this evolution, together with architectural choices and techniques aimed at improving the network energy efficiency (Green Networking). As a result, optimal dynamic resource allocation strategies should be readily available to support the current workload generated by applications at the required Quality of Service/Quality of Experience (QoS/QoE) levels, with minimum energy expenditure. In this framework, we briefly explore the above-mentioned paradigms, and describe their potential application in a couple of satellite networking related use cases, regarding traffic routing and gateway selection, and satellite swarms, respectively.

Keywords: Network flexibility · Network programmability · SDN · NFV · OpenFlow · Satellite networking

1 Introduction

Among other types of traffic, the Future Internet should support a very large number of heterogeneous user-led services, increased user mobility, machine-to-machine (M2M) communications, and multimedia flows with a massive presence of video. In order to face the challenges posed by the increased volume and differentiation of user-generated traffic, many Telecom operators believe that next-generation network devices and infrastructures should be more energy-efficient, scalable and flexible than those based on today's telecommunications equipment, along with a tighter integration among heterogeneous networking platforms (fixed, cellular wireless, and satellite). A possible promising solution to this problem seems to rely on extremely virtualized and "vertically" (across layers) optimized networks. At the same time, the interaction between the network and the computing infrastructure (user devices, datacenters and the cloud),

© ICST Institute for Computer Sciences, Social Informatics and Telecommunications Engineering 2016
I. Bisio (Ed.): PSATS 2014, LNICST 148, pp. 1–11, 2016.
DOI: 10.1007/978-3-319-47081-8_1

where applications reside, needs to be redesigned and integrated, with the aim of achieving greater use of mass standard Information Technology (IT), ease of programmability, flexibility in resource usage, and energy efficiency (goals actually pursued since a long time in the IT world, also by means of virtualization techniques). In all network segments (access, metro/transport and core), and across different networking infrastructures, this attitude, aiming at leveraging on IT progress, as well as achieving energy consumption proportional to the traffic load, is rapidly being adopted [1–3]. In this perspective, energy efficiency also plays a central role, and can be viewed as an indicator of the "health" of the overall computing and networking ecosystem. It reflects the extent of exploitation of computing, storage, and communications hardware capabilities to the degree needed to support the current workload generated by applications at the required Quality of Service/Quality of Experience (QoS/QoE) levels. Thus, flexibility and programmability of the network itself and of all other physical resources come naturally onto the scene as instruments that allow optimal dynamic resource allocation strategies to be really implemented in practice.

In this short note, we will explore the state of the art in energy-efficiency in various networking platforms, including the satellite segment, and the integration of green technologies in the framework of two emerging paradigms for network programmability and flexibility – Software Defined Networking (SDN) [4, 5] and Network Functions Virtualization (NFV) [6] – as a sustainable path toward the Future Internet.

2 Flexibility and Programmability in the Network

Bottlenecks in the networking infrastructure have been changing over time. Whereas one of the main bottlenecks once used to be bandwidth (still to be administered carefully in some cases, though), the increase in the capacity of transmission resources and processing speed, paralleled by an unprecedented increase in user-generated traffic, has brought forth other factors that were previously concealed. Among others, some relevant aspects are:

- The networking infrastructure makes use of a large variety of hardware appliances, dedicated to specific tasks, which typically are inflexible, energy-inefficient, unsuitable to sustain reduced Time to Market of new services;
- The so-called "ossification" of the TCP/IP architectural paradigm and protocols – implemented most of the time on proprietary components – is hindering the capability to host evolutions/integrations in the standards;
- The efficient (in terms of resource usage) management and control of flows, be they user-generated or stemming from aggregation, has become increasingly complex in a purely packet-oriented transport and routing environment.

Then, as one of the main tasks of the network is allocating resources, a natural question is how to provide architectural frameworks capable of efficiently supporting algorithms and techniques that can make this task more dynamic, performance-optimized and cost-effective. Current keywords in this respect are Flexibility, Programmability, and Energy-Efficiency. SDN and NFV aim at addressing the first two. We do not enter any details here (among others, see [4–6]), but only note some essentials. By decoupling the

Control Plane and the Data (Forwarding) Plane of devices, SDN allows a more centralized vision to set the rules for handling flows in the network, by means of specific protocols for the interaction between the controller and the devices under its supervision. OpenFlow is the most well known and widespread of such protocols and a paradigmatic example. It allows setting up, updating and modifying entries in a flow table on each forwarding device, by establishing matching rules, prescribing actions, managing counters and collecting statistics. On the other hand, NFV leverages "...standard IT virtualization technology to consolidate many network equipment types onto industry standard high volume servers, switches and storage, which could be located in Datacentres, Network Nodes and in the end user premises" [6]. It fosters improved equipment consolidation, reduced time-to-market, single platform for multiple applications, users and tenants, improved scalability, multiple open eco-systems; it exploits economy of scale of the IT industry (approximately 9.5 million servers shipped in 2011 against approximately 1.5 million routers). NFV requires swift I/O performance between the physical network interfaces of the hardware and the software user-plane in the virtual functions, to enable sufficiently fast processing, and a well-integrated network management and cloud orchestration system, to benefit from the advantages of dynamic resource allocation and to ensure a smooth operation of the NFV-enabled networks [3]. SDN is not a requirement for NFV, but NFV can benefit from being deployed in conjunction with SDN. Some examples of this integration are provided in [3], also in relation to energy-efficiency, which will be the subject of the next Section. For instance, an SDN switch could be used to selectively redirect a portion of the production traffic to a server running virtualized network functions. This way the server and functions do not need to cope with all production traffic, but only with the relevant flows. The SDN-enabled virtual switch running inside the server's hypervisor can dynamically redirect traffic flows transparently to an individual network function or to a chain of network functions. This enables a very flexible operation and network management, as functions can be plugged in and out of the service chain at runtime [3]. As the main focus of these notes is on the relevance of these architectural paradigms and techniques in the context of satellite networking, we can remark explicitly that, among functionalities that would lend themselves to such treatment, we might include many of those typically delegated to Performance Enhancing Proxies (PEPs), a kind of middlebox quite frequently encountered in satellite communications.

Essentially, with the adoption of these two paradigms, the premises are there for a – technically and operationally – easier way to more sophisticated and informationally richer network control (quasi-centralized/hierarchical vs. distributed) and network management. The latter may exhibit a tighter integration with control strategies, and closer operational tools, with perhaps the main differentiation coming in terms of time scales of the physical phenomena being addressed. In our opinion, the technological setting brought forth by the new paradigms enables the application of the philosophy that was at the basis of some of the early works on hierarchical multi-level and multi-layer control concepts, both in the industrial control and networking areas [7–9], to an unprecedented extent.

3 Energy Efficiency

How does all this interact with network energy-efficiency? As a matter of fact, making the network energy-efficient ("Green") cannot ignore QoS/QoE requirements. At the same time, much higher flexibility, as well as enhanced control and management capabilities, are required to effectively deal with the performance/power consumption tradeoff, once the new dimension of energy-awareness is taken into account in all phases of network design and operation. The enhanced control and management capabilities and their tighter integration offered by the application of SDN and NFV concepts go exactly in that direction.

The reasons that drive the efforts toward "greening" the network are well known [10, 11], and the impact of green networking on cutting the power consumption and Operational Expenditure (OPEX) is non negligible [12]. Again without entering too many details, we are particularly interested here in recalling the potential of the group of techniques known as Dynamic Adaptation, where two among the typical control actions that can be applied are Low Power Idle (LPI) and Adaptive Rate (AR), consisting of the modulation of "energy operating states" in the absence and presence of traffic, respectively [11]. Their effect can be summarized in the "power profile" of energy-aware components of network devices, i.e., in the characterization of the power consumption as a function of the traffic load [12]. In terms of QoS, the difference among operating states lies essentially in the wakeup times from "sleeping modes" for LPI (where lower power consumption implies longer wakeup time) and in different operating frequencies and/or applied voltage for AR (which affects processing capacity). Therefore, there is a natural tradeoff between power and performance, which can be optimized for different values of traffic load. Given a certain number of operating states, there are then basically two different kinds of control strategies to perform Dynamic Adaptation: (i) entering a certain LPI configuration when no packet is present to be processed in a specific component of the device and exiting to a certain active configuration upon packet arrivals (which can be "sensed" in different ways); (ii) choosing the idle and operating configurations in order to optimize some long-term figure of merit (e.g., minimize average delay, average energy consumption, or a combination thereof), while at the same time respecting some given constraints on the same quantities. In the first case, control is effected at the packet level; the strategy is dynamic and based on instantaneous local information (presence or absence of packets). In the second case the control can be based on parametric optimization, typically relying on information acquired over a relatively long term (e.g., in time scales of minutes, possibly comparable to flow dynamics – anyway several orders of magnitude greater than the time scales of packet dynamics) and typically related to long-term traffic statistics (average intensity, average burst lengths, etc.). The parametric optimization with respect to energy configurations can be combined with other traffic-load related optimizations, like load balancing in multi-core device architectures [13].

It is worth noting that optimization techniques at different time scales require some form of modeling of the dynamics of the system under control. In this respect, models based on "classical" queuing theory [13, 14] lend themselves to performance analysis or parametric optimization for adaptive control and management policies over the longer time scales (with respect to queueing dynamics). The already cited examples are in packet processing engines at routers' line cards [13] and in Green Ethernet transmission modules [14]. On the other hand, fluid models suitable for real-time control can be derived from the classical queueing equations (we recall here the very interesting approach pursued in [15]), or even from simpler, measurement-based, stochastic continuous fluid approximations [16]. In [15], optimal dynamic control strategies were applied upon fluid models derived from the classical queueing theory approach, but capable of describing the dynamic evolution of average quantities of interest (e.g., queue lengths). In our opinion, it would be worth revisiting the approach in the light of the new power consumption/performance tradeoff.

The above-mentioned models and techniques are suitable for Local Control Policies (LCPs), to be applied at the device level. However, it is also important to be able to establish energy-aware Traffic Engineering and routing policies at a "global" level (i.e., regarding a whole network domain), residing in the Control Plane and typically acting on flows, which we can refer to as Network Control Policies (LCPs). These have been considered in the recent literature, for instance in [17–20], also in relation with SDN capabilities [20]. In this respect, a relevant issue concerns the interaction between LCPs and NCPs, and the way to expose energy-aware capabilities, energy-profiles and energy-related parameters affecting QoS (e.g., wakeup delays) toward the Control Plane. A significant step in this direction has been achieved through the definition of the Green Abstraction Layer (GAL) [21, 22], now an ETSI standard [23], which allows summarizing the essential characteristics that are needed to implement energy-aware NCPs and to possibly modify device-level parameter settings accordingly.

Whereas most of the recent work cited so far was implemented in the framework of the ECONET project [24], which was devoted to energy-efficiency in the fixed network, it is worth pointing out that very similar situations in which Dynamic Adaptation strategies find useful applications are encountered also in the wireless environment [25, 26] and in datacenters [27, 28].

4 Satellites in a Green and Flexible Heterogeneous Networking Environment

A recent survey on energy-efficiency in satellite networking is that of Alagöz and Gür [29]. They discuss aspects related to the device level (terminal/earth station/satellite payload) regarding security and energy efficiency, energy constraints in the airborne platform, integration with the terrestrial segment, mobile terminals, as well as networking aspects, particularly in the context of hybrid heterogeneous networks, with the satellite playing the role of relay between various access networks and the core. They also explore emerging factors such as dynamic spectrum access and cognitive radio, cross-layer design, integration of space/terrestrial networks, Smart Grid support, emergency communications, and the Interplanetary Internet. Among some additional recent works related to energy-efficient satellite communications that appeared after the survey we can cite [30–33]. Reference [33] is related to one of the two exemplary topics we will briefly discuss in the following, and it applies what appears to be a very promising optimal control technique, based on Lyapunov optimization [34].

Here we consider two different satellite environments in their relation with flexible and green networking: (i) High Throughput Satellite (HTS) systems (at Terabit/s capacity) [35]; (ii) Nano-satellite networks (or, more generally, satellite swarms) [36].

4.1 HTS Scenario

HTS systems operate in Ka band to the users, but the scarcity of the available spectrum pushes to the use of the Q/V (40/50 GHz) bands for the gateways [37]. At these high bandwidths, where rain attenuation can produce particularly deep fading, gateway diversity is adopted to ensure the required feeder link availability [38, 39]. In essence, when each user is assigned to a pool of gateways (so-called Smart Gateways), a switching decision must be taken whenever the gateway serving the user experiences deep fading, to reroute the traffic to another unfaded gateway. Apart from the different architectural choices and ways to achieve the goal, gateway cooperation is required to efficiently obtain the desired availability level at a reasonable cost. Handover decisions should be taken at the Network Control Center (NCC), where channel state information from all the gateways should be conveyed.

At the same time, in integrated satellite-terrestrial architectures such as that envisioned by the BATS (Broadband Access via integrated Terrestrial & Satellite systems) project [35], Intelligent Network Gateways (INGs), as well as their user-side counterparts Intelligent User Gateways (IUGs), will be required to take routing decisions on traffic flows, on the basis of QoS/QoE requirements.

Then, let us recall the SDN and energy-aware scenario sketched in Sects. 2 and 3 above, and consider a situation were proper enhancement to OpenFlow allows taking advantage of the information conveyed through the GAL [20]. We can then imagine to have SDN-enabled network nodes (possibly a subset of them [40]), capable of executing power management primitives (e.g., Dynamic Adaptation, Sleeping/Standby) and associated LCPs, and an SDN Control Plane with an Orchestrator/NCC (that can reside in a cloud) in charge of implementing NCPs. SDN network nodes can include Smart Satellite Gateways, either directly or indirectly (through the SDN-enabled upstream router). Each interaction between the NCPs and the LCPs is performed according to the OpenFlow Specification.

Then, we can envisage a situation as depicted in Fig. 1, where incoming traffic is (dynamically at the flow level) directed to terrestrial or satellite paths according to joint Energy Efficiency and QoS/QoE performance indexes, and decisions are taken (dynamically with respect to channel outage conditions) on redirecting flows (or re-adjusting their balance [41]) among satellite gateways. We do not maintain the necessity of SDN for the implementation of such scenario (nor its straightforward feasibility); however, the architectural implications, the possible solutions, the required protocol extensions and the performance evaluation are certainly worth investigating.

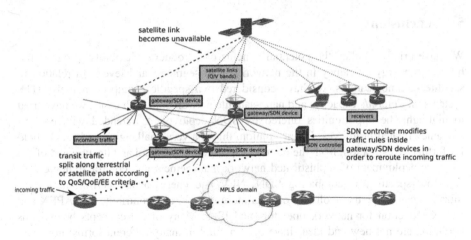

Fig. 1. HTS scenario integrated with SDN.

4.2 Satellite Swarms

There is a recent growing interest in this area, owing to the continuous development of the Internet of Things and to the desire to overcome the digital divide [36, 42], fostered by the relatively low cost of such solutions as compared to the traditional non-geostationary (NGEO) ones. Operating according to a Delay Tolerant Networking (DTN) paradigm [43] is practically a must here, and we should note the "intrinsic" energy-efficient operating characteristics of DTN. By forming a store-and-forward overlay network at the Bundle Layer [44], DTN performs grouping of smaller messages into larger aggregations, which can then be scheduled for transmission opportunities. In terms of exploiting the smart-sleeping techniques that constitute a category of methodologies for green networking, this kind of operating characteristics tends to increase the overall energy-efficiency of the system. Indeed – though operating at the packet level – one of the earliest proposed strategies to exploit smart sleeping and adaptive rate techniques has been the so-called "buffer and burst" [45], and "packet coalescing" has been suggested in connection with the Green Ethernet [46]. Forwarding decisions could then be taken at the bundle layer with attention to link/node availability and delay, but also to energy efficiency.

Recent work in this area [47] has taken into consideration the dynamic "hot spot" selection, where hot spots here play the role of small gateways that upload bundles to the satellites, which will then forward them to "cold spots" connecting users in rural or secluded areas. Here again, providing SDN capabilities to the hot spots and to the central node of the nano-satellite constellation is worth investigating, from the architectural and performance evaluation points of view.

5 Conclusions

We have briefly recalled the potential benefits of introducing flexibility, programmability and energy efficiency in the network, at all segments and levels. In relation to satellite communications, we have considered two specific examples, namely, HTS systems (at Terabit/s capacity) and nano-satellite networks. In both cases, we have tried to highlight the opportunities offered by SDN deployment, extended with energy-efficiency related primitives. In our opinion, this is a very challenging and timely field for further investigation, from the point of view of both protocol architecture and of the effective deployment of sophisticated network management and control strategies.

More specifically, combining SDN, NFV and energy-aware performance optimization can shape the evolution of the Future Internet and contribute to CAPEX and OPEX reduction for network operators and ISPs. Many of the concepts behind this evolution are not new and ideas have been around in many different forms; however, current advances in technology make them feasible. Sophisticated control/management techniques can be realistically deployed – both at the network edge and inside the network – to dynamically shape the allocation of resources and relocate applications and network functionalities, trading off QoS/QoE and energy at multiple granularity levels. Satellite networking does fit in this scenario as a relevant component, by:

- Providing energy efficient by-passes in the backhaul;
- Dynamically diverting flows, while preserving QoS/QoE requirements;
- Benefiting of increased flexibility in resource allocation to compensate fading in Q/V band smart diversity for Terabit/s speeds;
- Integrating with terrestrial networks;
- Adding energy efficient solutions in the access network for rural areas (nano-satellites and DTN);
- Benefiting of virtualization in the flexible implementation of related functionalities (PEP, optimization strategies in the cloud, …);
- Participating in consolidation of flows over a limited number of paths where possible.

Further research activities are needed for the full development of a large spectrum of possibilities.

References

1. Manzalini, A.: Future edge ICT networks. IEEE ComSoc MMTC E-lett. **7**(7), 1–4 (2012)
2. Manzalini, A., Minerva, R., Callegati, F., Cerroni, W., Campi, A.: Clouds of virtual machines in edge networks. IEEE Commun. Mag. **51**(7), 53–70 (2013)
3. Jarschel, M., Hoßfeld, T., Davoli, F., Bolla, R., Bruschi, R., Carrega, A.: SDN-enabled energy-efficient network management. In: Samdanis, K., Rost, P., Maeder, A., Meo, M., Verikoukis, C. (eds.) Green Communications: Principles, Concepts and Practice. Wiley, Chichester (2015)

4. Software-Defined Networking: The New Norm for Networks, Open Networking Foundation (ONF) White Paper (2012)
5. Nunes, B.A.A., Mendonça, M., Nguyen, X.-N., Obraczka, K., Turletti, T.: A survey of software-defined networking: past, present, and future of programmable networks. IEEE Commun. Surv. Tutorials **16**(3), 1617–1634 (2014)
6. Network Functions Virtualisation – Introductory White Paper. SDN and OpenFlow World Congress, Darmstadt, Germany (2012)
7. Lefkowitz, I.: Multilevel approach applied to control system design. J. Basic Eng. Trans. ASME **88**(Ser. B2), 392–398 (1966)
8. Findeisen, W., Bailey, F.N., Brdyś, M., Malinowski, K., Tatjewski, P., Woźniak, A.: Control and Coordination in Hierarchical Systems. Wiley, Chichester (1980)
9. Hui, J.Y.: Resource allocation for broadband networks. IEEE J. Select. Areas Commun. **6**(9), 1598–1608 (1988)
10. Global e-Sustainability Initiative (GeSI) Report. SMARTer2020: The Role of ICT in Driving a Sustainable Future. http://gesi.org/SMARTer2020
11. Bolla, R., Bruschi, R., Davoli, F., Cucchietti, F.: Energy efficiency in the future internet: a survey of existing approaches and trends in energy-aware fixed network infrastructures. IEEE Commun. Surv. Tutorials **13**(2), 223–244 (2011)
12. Bolla, R., Bruschi, R., Carrega, A., Davoli, F., Suino, D., Vassilakis, C., Zafeiropoulos, A.: Cutting the energy bills of internet service providers and telecoms through power management: an impact analysis. Comput. Netw. **56**(10), 2320–2342 (2012)
13. Bolla, R., Bruschi, R., Carrega, A., Davoli, F.: Green networking with packet processing engines: modeling and optimization. IEEE/ACM Trans. Netw. **22**(1), 110–123 (2014)
14. Bolla, R., Bruschi, R., Carrega, A., Davoli, F., Lago, P.: A closed-form model for the IEEE 802.3az network and power performance. IEEE J. Select. Areas Commun. **32**(1), 16–27 (2014)
15. Filipiak, J.: Modelling and Control of Dynamic Flows in Communication Networks. Springer, Berlin (1988)
16. Bruschi, R., Davoli, F., Mongelli, M.: Adaptive frequency control of packet processing engines in telecommunication networks. IEEE Commun. Lett. **18**(7), 1135–1138 (2014)
17. Bolla, R., Bruschi, R., Cianfrani, A., Listanti, M.: Enabling backbone networks to sleep. IEEE Netw. **25**(2), 26–31 (2011)
18. Bianzino, A.P., Chiaraviglio, L., Mellia, M., Rougier, J.-L.: GRiDA: green distributed algorithm for energy-efficient IP backbone networks. Comput. Netw. **56**(14), 3219–3232 (2012)
19. Niewiadomska-Szynkiewicz, E., Sikora, A., Arabas, P., Kołodziej, J.: Control system for reducing energy consumption in backbone computer networks. Concurrency Comput. Pract. Exp. **25**(12), 1738–1754 (2013)
20. Bolla, R., Bruschi, R., Davoli, F., Lombardo, A., Lombardo, C., Morabito, G., Riccobene, V.: Green extension of OpenFlow. In: ITC 26 Workshop on Energy-efficiency, Programmability, Flexibility and Integration in Future Network Architectures (EPFI), Karlskrona, Sweden (2014)
21. Bolla, R., Bruschi, R., Davoli, F., Di Gregorio, L., Donadio, P., Fialho, L., Collier, M., Lombardo, A., Reforgiato Recupero, D., Szemethy, T.: The green abstraction layer: a standard power management interface for next-generation network devices. IEEE Internet Comput. **17**(2), 82–86 (2013)
22. Bolla, R., Bruschi, R., Davoli, F., Donadio, P., Fialho, L., Collier, M., Lombardo, A., Reforgiato, D., Riccobene, V., Szemethy, T.: A northbound interface for power management in next generation network devices. IEEE Commun. Mag. **52**(1), 149–157 (2014)

23. ETSI Standard ES 203 237 V1.1.1, Environmental Engineering (EE); Green Abstraction Layer (GAL); Power Management Capabilities of the Future Energy Telecommunication Fixed Network Nodes, ETSI (2014)
24. http://www.econet-project.eu
25. Correia, L.M., Zeller, D., Blume, O., Ferling, D., Jading, Y., Gódor, I., Auer, G., Van der Perre, L.: Challenges and enabling technologies for energy aware mobile radio networks. IEEE Commun. Mag. **48**(11), 66–72 (2010)
26. Scalia, L., Biermann, T., Choi, C., Kozu, K., Kellerer, W.: Power-efficient mobile backhaul design for CoMP support in future wireless access systems. In: IEEE INFOCOM 2011 Workshop on Green Communications and Networking, Shanghai, China, pp. 253–258 (2011)
27. Baliga, J., Ayre, R.W.A., Hinton, K., Tucker, R.S.: Green cloud computing: balancing energy in processing, storage, and transport. Proc. IEEE **99**(1), 149–167 (2011)
28. Liu, J., Zhao, F., Liu, X., He, W.: Challenges towards elastic power management in internet data centers. In: IEEE International Conference on Distributed Computing Systems Workshops, Los Alamitos, CA, pp. 65–72 (2009)
29. Alagöz, F., Gür, G.: Energy efficiency and satellite networking: a holistic overview. Proc. IEEE **99**(11), 1954–1979 (2011)
30. Kandeepan, S., Rasheed, T., Reisenfeld, S.: Energy efficient cooperative HAP-terrestrial communication systems. In: Giambene, G., Sacchi, C. (eds.) PSATS 2011. LNICST, vol. 71, pp. 151–164. Springer, Heidelberg (2011)
31. Alagöz, F., Gür, G.: Energy-efficient layered content dissemination for multi-mode mobile devices. In: 2013 First International Black Sea Conference on Communications and Networking (BlackSeaCom), Batumi, Georgia, pp. 243–246 (2013)
32. Tang, Z., Wu, C., Feng, Z., Zhao, B., Yu, W.: Improving availability through energy-saving optimization in LEO satellite networks. In: Mahendra, M.S., et al. (eds.) ICT-EurAsia 2014. LNCS, vol. 8407, pp. 680–689. Springer, Heidelberg (2014)
33. An, Y., Li, J., Fang, W., Wang, B., Guo, Q., Li, J., Li, X., Du, X.: EESE: Energy-efficient communication between satellite swarms and earth stations. In: 16th IEEE International Conference on Advanced Communication Technology (ICACT 2014), PyeongChang, Korea, pp. 845–850 (2014)
34. Neely, M.J.: Stochastic Network Optimization with Application to Communication and Queueing Systems. Morgan & Claypool, San Rafael (2010)
35. Pérez-Trufero, J., Watts, S., Peters, G., Evans, B., Fesquet, T., Dervin, M.: Broadband access via integrated terrestrial and satellite systems (BATS). In: 19th Ka and Broadband Communications, Navigation and Earth Observation Conference, Florence, Italy, pp. 27–34 (2013)
36. Burleigh, S.: Nanosatellites for universal network access. In: 2013 ACM MobiCom Workshop on Lowest Cost Denominator Networking for Universal Access (LCDNet 2013), Miami, FL, pp. 33–34 (2013)
37. Kyrgiazos, A., Evans, B., Thompson, P., Mathiopoulos, P.T., Papaharalabos, S.: A Terabit/second satellite system for European broadband access: a feasibility study. Int. J. Satell. Commun. Netw. **32**(2), 63–92 (2014)
38. Jeannin, N., Castanet, L., Radzik, J., Bousquet, M., Evans, B., Thompson, P.: Smart gateways for Terabit/s satellite. Int. J. Satell. Commun. Netw. **32**(2), 93–106 (2014)
39. Gharanjik, A., Rama Rao, B.S.M., Arapoglou, P.-D., Ottersten, B.: Gateway switching in Q/V band satellite feeder links. IEEE Commun. Lett. **17**(7), 1384–1387 (2013)
40. Agarwal, S., Kodialam, M., Lakshman, T.V.: Traffic engineering in software defined networks. In: IEEE INFOCOM 2013, Torino, Italy, pp. 2211–2219 (2013)

41. Kyrgiazos, A., Thompson, P., Evans, B.: Gateway diversity via flexible resource allocation in a multibeam SS-TDMA system. IEEE Commun. Lett. **17**(9), 1762–1765 (2013)
42. https://www.outernet.is
43. http://www.dtnrg.org/wiki/Home
44. Scott, K., Burleigh, S.: Bundle Protocol Specification. RFC 5050, IETF (2007)
45. Nedevschi, S., Popa, L., Iannaccone, G., Ratnasamy, S., Wetherall, D.: Reducing network energy consumption via sleeping and rate-adaptation. In: 5th USENIX Symposium on Networked Systems Design and Implementation (NSDI 2008), San Francisco, CA, pp. 323–336 (2008)
46. Christensen, K., Reviriego, P., Nordman, B., Bennett, M., Mostowfi, M., Maestro, J.A.: IEEE 802.3az: the road to energy efficient ethernet. IEEE Commun. Mag. **48**(11), 50–56 (2010)
47. Cello, M., Marchese, M., Patrone, F.: Hot spot selection in rural access nanosatellite networks. In: ACM CHANTS 2014, Maui, HI (2014)

Extended Future Internet: An IP Pervasive Network Including Interplanetary Communication?

Mario Marchese[(✉)]

Department of Electrical, Electronic and Telecommunications Engineering,
and Naval Architecture (DITEN), University of Genova,
Via Opera Pia 13, 16145 Genoa, Italy
mario.marchese@unige.it

Abstract. Starting from the evolution of Internet, this paper addresses the concept of pervasive computing whose aim is to create a pervasive network of heterogeneous devices which communicate data with each other and with other networking devices in a seamless way through heterogeneous network portions. This operative framework is also called Future Internet. Extending the idea of pervasive computing to interplanetary and other challenging links implies adding to the classical problems of pervasive communications such as quality of service, mobility and security, peculiarities such as intermittent connectivity, disruptive links, large and variable delays, and high bit error rates which are currently tackled through the paradigm of Delay and Disruption Tolerant Networking (DTNs). Satellite systems used to connect isolated and rural areas have already to cope with a series of challenges that are magnified in space communications characterized by huge distances among network nodes. At the same time, a space communication system must be reliable over time and the importance of enabling Internet-like communications with space vehicles (as well as with rural areas) is increasing, making the concept of extended Future Internet of practical importance. This paper will discuss this challenging issue.

Keywords: Internet · Pervasive communications · Future internet · Satellite communications · Delay and Disruption Tolerant Networking (DTN)

1 Introduction: Internet Evolution

The first step towards Future Internet is having a widespread diffusion of the Internet throughout the world. Table 1 reports the estimated population at the end of 2013 and the estimated number of Internet users at the end of 2013 and 2000 structured for world regions, showing also the world average. All data in this section are taken from [1].

Figure 1 shows the estimated Internet penetration rate (i.e. the percentage of estimated Internet users over the estimated population) in Dec. 2013 for each world region and for the world average. Penetration rate in North America is astonishing, and satisfying data are estimated for Europe and Oceania/Australia. Penetration rates in Middle East, Latin America/Caribbean, and, in particular, in Asia and Africa show that much work must still be done to fill the digital divide among world regions but, if, on

© ICST Institute for Computer Sciences, Social Informatics and Telecommunications Engineering 2016
I. Bisio (Ed.): PSATS 2014, LNICST 148, pp. 12–21, 2016.
DOI: 10.1007/978-3-319-47081-8_2

one hand, this is a negative factor, on the other hand, the analysis of data evidences both the huge growth of Internet users in Asia, Middle East, Latin America/Caribbean, and Asia from 2000 to 2014, clear in Fig. 2, and the great potential of Asia, Africa, and Latin America/Caribbean due to the amount of population in these world regions. Figure 3, which shows the percentage of Internet users in the world distributed by world regions in Dec. 2013, may help evidence this last aspect: even if the estimated penetration rate in Asia is under 32 % for now, the number of estimated Internet users in this region is above 1.2 billions, which represent more than 45 % of the Internet users in the world. This fact, associated to an impressive growth of more than 1000 % in these last 13 years, allows envisaging a key role of Asia in Future Internet. Similar observations may be reported for Africa, which has a penetration rate of about 21 % but a 2000–2013 growth higher than 5200 % and a global population above 1 billion.

Table 1. Data about estimated population and estimated Internet users structured for world regions.

World regions	Estimated population, Dec. 31, 2013	Estimated internet users, Dec. 31, 2000	Estimated internet users, Dec. 31, 2013
Africa	1,125,721,038.00	4,514,400.00	240,146,482.00
Asia	3,996,408,007.00	114,304,000.00	1,265,143,702.00
Europe	825,802,657.00	105,096,093.00	566,261,317.00 ·
Middle East	231,062,860.00	3,284,800.00	103,829,614.00
North America	353,860,227.00	108,096,800.00	300,287,577.00
Latin America/Caribbean	612,279,181.00	18,068,919.00	302,006,016.00
Oceania/Australia	36,724,649.00	7,620,480.00	24,804,226.00
WORLD TOTAL	7,181,858,619.00	360,985,492.00	2,802,478,934.00

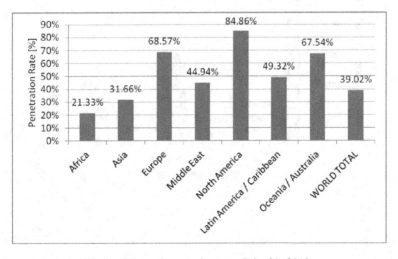

Fig. 1. Internet penetration rate, Dec. 31, 2013.

Concerning the mentioned digital divide, it has many complex motivations [2] including the following figures: temporal (having time to use digital media), material (possession and income), mental (technical ability and motivation), social (having a social network to assist in using digital media), and cultural (status and liking of being in the world of digital media), but one of the reasons is that a large amount of people lives in countries or in remote areas which do not have a suitable telecommunication infrastructure. The costs needed to connect these areas by using cables and common infrastructures are very high, in particular if compared with economic benefits. Satellite communications constitute a strategic sector for service provision in remote and low density population areas, as well as for aeronautical services, disaster prediction and relief, safety for critical users, search and rescue, data transmission for maritime environment, aviation and trains, and crisis management. The challenge is if satellite technology can fill the digital divide at service cost, reliability and quality comparable to terrestrial solutions. Actually, current satellite technologies require high costs in the construction, launch and maintenance, but nanosatellites [3] have been recently proposed as a cost-effective solution to extend the network access in rural and remote areas. Rural and/or disconnected areas can be connected through local gateways that will communicate with the nanosatellite constellation. The availability of the connection with nanosatellites is not permanently guaranteed and it deserves a dedicated solution, called DTN – Delay and Disruption Tolerant Networking, discussed in the remainder of the paper.

Given these data, is pervasive computing feasible? Next section provides more detail about this paradigm and about its evolution to Future Internet.

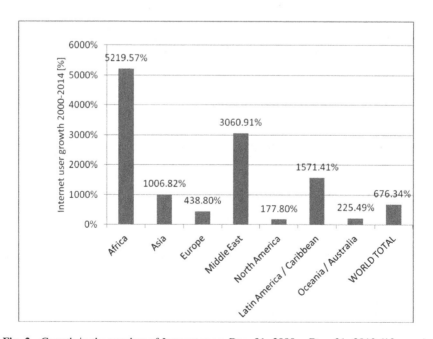

Fig. 2. Growth in the number of Internet users Dec. 31, 2000 – Dec. 31, 2013 (13 years).

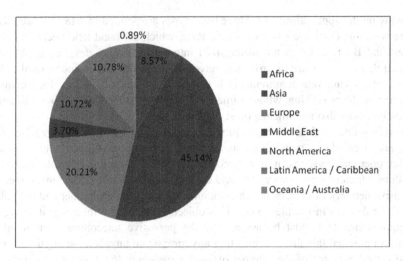

Fig. 3. Percentage of Internet users in the world distributed by world regions, Dec. 31, 2013.

2   From Pervasive Computing to Future Internet

The paradigm of pervasive computing, also called ubiquitous computing, is a model of human-machine interaction where computing and processing power is totally integrated in everyday objects and activities. These objects can also communicate with each other and with other components so forming a pervasive/ubiquitous communication network. The idea, perfectly focused by [4], is sensing physical quantities, which presents a wide set of input modalities (vibrations, heat, light, pressure, magnetic fields,…), through sensors and transmit them by using seamless communication networks for information, decision, and control aim. Historically the concept of ubiquitous computing and net-working was introduced by Mark Weiser and is contained in the paper [5] that envisages a world where sensors and digital information are integral part of people everyday life. The imagine that comes from that is the imagine of a person totally immersed within a telecommunication network who sends and receives digital information from the sur-rounding physical world and who interacts with it also unconsciously. The alarm clock asks about the will of drinking a coffee and activates the coffee machine in case of positive vocal answer; electronic trails reveal the presence of neighbours; evidencing some lines by a special pen in a newspaper it is sufficient to send these lines to your office for further elaboration. All these examples are taken from [5] but others may be created: the refrigerator gives indication about the status of the food; the washing machine and the heater may be switched on remotely; the car engine ignition may be switched on automatically when the owner is approaching, and so on. Obviously examples are not limited to home applications but extend to all environments where monitoring and connecting physical world is important: civil protection, transportation, military, underwater, space monitoring and communications, among the others. As written in [4], "We foresee thousands of devices embedded in the civil infrastructure (buildings, bridges, water ways, highways, and protected regions) to monitor structural health and detect crucial events". Used embedded devices change their dimension depending on the application field. Three basic types are defined by Mark Weiser: Tabs that are wearable centimetre sized devices, Pads, which are hand-held decimetre-sized devices, and Boards, which are meter sized interactive display devices. In Weiser's vision all these devices are macro-sized, have a planar form and include visual output displays. Removing this requirements brings to new sets of devices for pervasive computing and networking whose dimension can be reduced down to millimetres, micrometers, and also nanometres (dust devices).

Interdisciplinary advances are required to innovate in the field of pervasive computing and networking: new communication and networking solutions, new and less complex operating systems, miniaturized memorization capacity, innovative decision algorithms, efficient signal processing and context aware solutions. The aim is to create a pervasive network of devices which communicate data with each other and with other networking devices in seamless way. This objective imposes a meaningful change in the requirements that must be assured by the pervasive telecommunication infrastructure. In practice the aim is connecting anything, from anyplace, at anytime. These are the three keywords of the Internet of Things paradigm [6], born independently of pervasive networking but now strictly connected to it. At least from the viewpoint of telecommunications the concepts of Pervasive Networks and Internet of Things are not distinguishable. Internet of Things refers to a network of objects to which has been given an electronic identity and some active features. Connecting the objects to each other and to other systems creates a pervasive network.

A pervasive network, so, is a telecommunication network composed of heterogeneous devices, differentiated for sizes, dynamics, and functions; and of heterogeneous communication solutions, ranging [7] from ADSL (Asymmetric Digital Subscriber Line) to DOCSIS (Data Over Cable Service Internet Specification); from fiber optic to PLC (Power Line Communication); from WiFi and its set of standards 802.11 dedicated to Wireless Local Area Networks – WLANs to WiMax, implemented through the 802.16 family, and LTE (Long Term Evolution), both suitable for the delivery of last mile wireless broadband access and to connect WiFi hotspots; from Bluetooth, acting over short ranges, to satellite solutions for planetary connections. Mentioned communication components not only implement different technologies but also often apply different protocols.

Figure 4 shows an example of pervasive network where there are many sensors that take physical measures and must transmit them remotely both to a mobile processing laboratory located on a plane and to a central laboratory located in a building (headquarters). In one case, data acquired from sensors are transmitted to a mobile station located on a off-road vehicle and, from there, to a satellite earth station through a wireless link. Data are broadcast through the satellite to an aeronautical network and, from there, forwarded to headquarters. In the other case data from sensors are directly received by a satellite earth station through a proper ad-hoc network and forwarded to headquarters via satellite.

Different network portions are connected by devices, called Interconnection Gateways in Fig. 4, whose role is to create a quality of service – guaranteed seemless interconnection of networks that implement different technologies and protocols.

Additionally some communication links may be not available in some periods of time. For example, observing Fig. 4, the link connectivity between the mobile station on the off-road vehicle and the satellite earth station may be intermittent because of the position of the vehicle; also the aeronautical link may be intermittent due to the plane position. In this case it would be recommendable that interconnection gateways could store information up to connection availability. This feature is mandatory in inter-planetary and nanosatellite communications where intermittent links are a typical situation but may be very important also in other environments. Extending the idea of pervasive computing to interplanetary and other challenging links implies adding to the classical problems of pervasive communications such as quality of service, mobility and security, the peculiarities of interplanetary links such as intermittent connectivity, disruptive links, large and variable delays, and high bit error rates which are currently tackled through the paradigm of Delay and Disruption Tolerant Networking (DTNs). The idea is including within the pervasive IP network called Future Internet also interplanetary and challenging links, such as nanosatellites, connecting remote locations so creating an Extended Future Internet. An example is shown in Fig. 5. As in Fig. 4, some data measured remotely must be delivered to a data centre but, in this case, acquisition sensors are located on a remote planet and data centre on the Earth.

Satellite systems used to connect isolated and rural areas have to cope with a series of challenges such as long round trip times (RTTs); likelihood of data loss due to errors on the communication link; possible channel disruptions; and coverage issues at high latitudes and in challenging terrain. These problems are magnified in space communications characterized by huge distances among network nodes, extremely long delays, and intermittent connectivity. At the same time, a space communication system must be reliable over time, for example, due to the long duration of space missions, or due to the content of communications in rural areas. Moreover the importance of

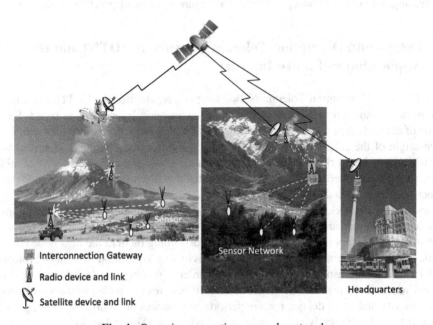

Fig. 4. Pervasive computing, example network.

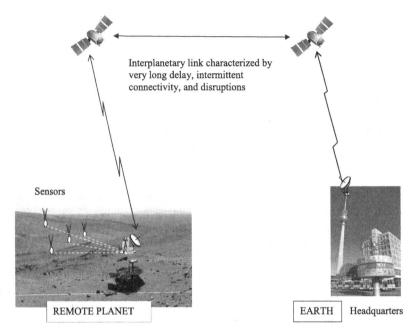

Fig. 5. Example of pervasive communication including long delay and intermittent connectivity.

enabling Internet-like communications with space vehicles as well as with rural areas is increasing, making the concept of extended Future Internet of practical importance.

3 Delay - and Disruption Tolerant Networking (DTN) and Its Application to Future Internet

The Delay and Disruption Tolerant Networking (DTN) architecture [8–11], introduces an overlay protocol that interfaces with either the transport layer or lower layers. Each node of the DTN architecture can store information for a long time before forwarding it. The origin of the DTN concept lies in a generalization of requirements identified for InterPlanetary Networking (IPN), where enormous latencies measured in tens of minutes, as well as limited and highly asymmetric bandwidth, must be faced. Nevertheless other scenarios, called "challenged networks", such as military tactical networking, sparse sensor networks, and networking in developing or otherwise communications-challenged regions can benefit from the DTN solution. Nodes on the path can provide the storage necessary for data in transit before forwarding them to the next node on the path. The contemporaneous end-to-end connectivity that Transmission Control Protocol (TCP) and other transport protocols require in order to reliably transfer application data is not required. In practice, in standard TCP/IP networks, which assume continuous connectivity and short delays, routers perform non-persistent (short-term) storage and information is persistently stored only at end nodes. In DTN networks information is

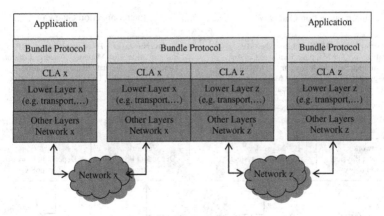

Fig. 6. DTN architecture.

persistently (long-term) stored at intermediate DTN nodes. This makes DTN much more robust against disruptions, disconnections, and node failures.

The Bundle Protocol (BP) is an implementation of the DTN architecture where the basic unit to transfer data is a Bundle, a message which carries application layer protocol data units, sender and destination names, and any additional data required for end-to-end delivery. The BP can interface with different lower layer protocols through convergence layer adapters (CLAs). CLAs for TCP, UDP, Licklider Transmission Protocol (LTP), Bluetooth, and raw Ethernet have been defined. Each DTN node can use the most suitable CLA to forward data. Generic DTN Architecture is shown in Fig. 6.

BP has important features such as: Custody Transfer, where an intermediate node can take custody of a bundle, relieving the original sender of the bundle which might never have the opportunity to retransmit the application data due to physical or power reasons; Proactive and Reactive Bundle Fragmentation, the former to tackle intermittent periodic connectivity when the amount of data that can be transferred is known a priori, the latter, which works ex post, when disruptions interrupt an ongoing bundle transfer; Late Binding, where, for example, when a bundle destination endpoint's identifier includes a Dynamic Name Server (DNS) name, only the CLA for the final DTN hop might have to resolve that DNS name to an IP address, while routing for earlier hops can be purely name based. Anyway, concerning the aim of this paper two are the BP features of main interest: (1) BP acts as an overlay layer and (2) can act as a long-term storage tool at intermediate nodes. These two features open the door to important applications of the DTN architecture, which:

- can be used as an alternative to PEP (Performance Enhancing Proxy) solutions,
- can be integrated within Interconnection Gateways that take care of quality of service – based internetworking among heterogeneous networks, as evidenced in the previous section,
- but can also store information and manage disruptions and long delays, if needed.

Generic idea is shown in Fig. 7.

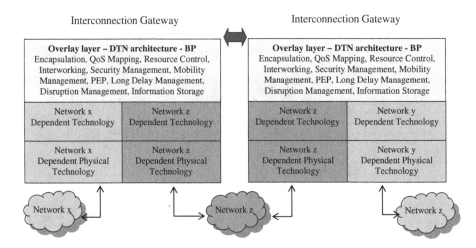

Fig. 7. DTN-based interconnection gateway.

4 Conclusions

This paper asks some basic questions: is pervasive communication extended to intermittent and disruptive links feasible and of practical interest or is it only an issue of academic investigation for now? To offer a possible answer the paper analyses the Internet evolution by showing the estimated number of Internet users at the end of 2013 structured for world regions and comparing these values with the same quantities at the end of 2000. Data concerning Asia and Africa show that much work must still be done to fill the digital divide among world regions but also show the huge growth of Internet users in Africa, Asia, Latin America/Caribbean, and Middle East from 2000 to 2014 and the great potential of these regions for the next future. This facts make the idea of connecting people and things from anyplace, at anytime, feasible. In the same time the importance of connecting rural areas, planets, and other remote locations characterized by intermittent and disruptive links makes the concept of Extended Future Internet a need. DTN offers a possible technical solution. So even if much research, in particular concerning modeling, routing, flow and congestion control, is still necessary to create a real Extended Future Internet, the challenge is worthwhile.

References

1. Internet World Stats, Usage and Population Statistics. http://www.internetworldstats.com/stats.htm
2. van Dijk, J.: The Evolution of the digital divide. In: Bus, J., et al. (eds.) Digital Enlightenment Yearbook 2012. IOS Press (2012)
3. Burleigh, S.: Nanosatellites for universal network access. In: Proceedings of the 2013 ACM MobiCom Workshop on Lowest Cost Denominator Networking for Universal Access (LCDNet 2013), Miami, FL, USA, September 2013

4. Estrin, D., Culler, D., Pister, K., Sukhatme, G.: Connecting the physical world with pervasive networks. Pervasive Comput. **1**(1), 59–69 (2002)
5. Weiser, M.: The computer for the 21st century. ACM SIGMOBILE Mob. Comput. Commun. Rev. Arch. **3**(3), 3–11 (1999). Special issue dedicated to Mark Weiser, reprinted, article first appeared in Scientific American, **265**(3), pp. 94–104, September 1991
6. Lahti, J.: The internet of things. In: Silverajan, B., (ed.) Pervasive Networks and Connectivity. Seminar Series on Special Topics in Networking, Spring 2008, pp. 58–64. Tampere University of Technology (2008). http://www.cs.tut.fi/~bilhanan/TLT2656_2008-Final.pdf
7. Lahteenmaki, E.: High speed network connectivity for homes and metropolitan areas. In: Silverajan, B., (ed.) Pervasive Networks and Connectivity. Seminar Series on Special Topics in Networking, Spring 2008, pp. 2–7. Tampere University of Technology (2008). http://www.cs.tut.fi/~bilhanan/TLT2656_2008-Final.pdf
8. Farrell, S.M.: Delay - and disruption-tolerant networking. IEEE Internet Comput. **13**(6), 82–87 (2009)
9. Cerf, V., Hooke, A., Torgerson, L., Durst, R., Scott, K., Fall, K., Weiss, H.: Delay-tolerant networking architecture. Internet RFC 4838, April 2007. http://www.rfc-editor.org/rfc/rfc4838.txt
10. Scott, K., Burleigh, S.: Bundle Protocol Specification, Internet RFC 5050, November 2007. http://www.rfc-editor.org/rfc/rfc5050.txt
11. Caini, C., Cruickshank, H., Farrell, S., Marchese, M.: Delay - and disruption-tolerant networking (DTN): an alternative solution for future satellite networking applications. Proc. IEEE **99**(11), 1980–1997 (2011). Invited Paper

A Fast Vision-Based Localization Algorithm for Spacecraft in Deep Space

Qingzhong Liang[(✉)], Guangjun Wang, Hui Li, Deze Zeng,
Yuanyuan Fan, and Chao Liu

School of Computer Science, China University of Geosciences, Wuhan 430074, China
{qzliang,gjwang,lihuicug,dzzeng,yyfan,liuchao}@cug.edu.cn

Abstract. Star light navigation can provide the current attitude and position of the spacecraft in deep space. However, the accuracy of stellar-inertial attitude determination is degraded because of star image smearing under high dynamic condition. To solve this problem, two key work, including accuracy star extraction and fast star identification, should be done. In this paper, we bring interpolation algorithm into contiguous area pixel searching for star extraction, and get sub-pixel coordinate information of the star points. In addition, a divisional method is proposed to improve star identification algorithm speed based on Hausdorff distance. The simulation results show that work not only has accuracy identification rate but also has better recognition speed. It was used successfully in the actual projects.

Keywords: Smearing image · Autonomous navigation · Star extraction

1 Introduction

Autonomous spacecraft navigation means the spacecraft can real-time determine its own position and attitude without any other support. The key factor in achieving autonomous navigation is accurate measurement of spacecraft attitude [1]. Currently, a new generation of CMOS star sensor is used for aircraft attitude measurement because of its high precision, none attitude cumulative error, fast fault recovery capability and intelligent [2]. It can provide accurate spacecraft flight attitude to a few arc-seconds without any prior knowledge. The star pattern recognition is one of the key technologies for spacecraft autonomous navigation based on star sensor. Many scholars are committed to this research, and proposed a number of algorithms. Nowadays, typical stellar identification algorithms used commonly include polygon angular distance matching algorithm, polygon angles matching algorithm, main star identification method proposed by Bezooijen, triangular matching algorithm, quadrilateral sky autonomous star identification algorithm, the sky autonomous grid algorithm, etc. [3–5]. Most of these algorithms complete recognition based on feature extraction. As a result, these complex algorithms are slow or need large storage and have poor anti-interference ability [6]. In addition, the star starlight images in the moving star

© ICST Institute for Computer Sciences, Social Informatics and Telecommunications Engineering 2016
I. Bisio (Ed.): PSATS 2014, LNICST 148, pp. 22–33, 2016.
DOI: 10.1007/978-3-319-47081-8_3

sensor will be stretched during the exposure time and lead to will lengthen and bring smearing. The smearing images reduce the centroid extraction accuracy, making a big decline in recognition accuracy decline.

In this paper, we aim at star centroid extraction in smearing images from moving star sensor. A extraction algorithm based on Gaussian curved interpolation is proposed to improve the centroid extraction accuracy. Secondly, we use the whole star database as a standard reference set, and consider extracted centroid data as a set to be recognized. Then, the minimum Hausdorff distance between two sets is determined to identify star location. At last, the divisional strategies for whole star database are proposed to improve the computational efficiency of the recognition algorithm.

2   Fast Vision-Based Localization Algorithm

Algorithm Overview. Because the star is considered to be at infinity, starlight can be seen as parallel to the light. In the inertial coordinate system, if the star sensor moving along a straight line, then the stars in the star sensor imaging position is fixed, which is similar to a static star sensor. However, when the star sensor rotates, stars' position detected in the star sensor will change and result in smearing. In Fig. 1, it shows a smearing image get from a moving star sensor.

Fig. 1. Case of smearing imaging.

Therefore, for deep space spacecraft location, the smearing images must be processed to extract the stellar centroid accurately. Then, it searches in whole star database with these stellar centroid data to get the current spacecraft location. The fast vision-based localization algorithm for spacecraft, proposed in this paper, consists of three steps, as shown in Fig. 2. First, Gaussian curved interpolation method for extraction of star centroid is used to obtain sub-pixel star centroid location information. Then, according to the moving state of the aircraft, we cut apart the whole star database, and match star centroid information in the divisional database to identification. Finally, the location of the aircraft in the whole star pattern is output as a result.

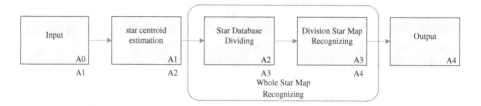

Fig. 2. Flow graph of fast vision-based localization algorithm.

2.1 Star Centroid Estimation Algorithm Based on Gauss Curved Fitting

Due to limitations of star sensor resolution, it is difficult to obtain high-precision stellar position from the star sensor image. Thus, there is a certain precision error in the extracted star pattern position. Set star sensor FOV (Field of View) of 100×100, the star sensor has a resolution of 1024×1024, the star sensor angular resolution is approximately 36", the error of the extracted star pattern position is also close to 36". Obviously, the error of extracting star pattern does not contribute to the correct rate of star pattern recognition, but also affect navigation accuracy. Taking into account the scattering of the lens, the imaging results in stellar star sensor should be a stellar position as the center of the spot. Because the star is a point light source, under normal circumstances the brightness of spots are represented by the point spread function, energy distribution can be approximated as a Gaussian surface, and the brightness decreases as quickly away from the center position. Considering the spot size is not large, and the point spread function of the specific parameter is difficult to determine. To solve this problem, the paper studies the Gaussian surface interpolation method to obtain analytic recursive Gaussian surface parameters.

As shown in Fig. 3, Set $p_0(x, y)$ is the maximum position of stars resulting from the star sensor images, coordinates (x', y'), its four adjacent gray values of $p_1(x1, y)$, $p_2(x2, y)$, $p_3(x, y1)$, $p_4(x, y2)$. Pixel $p_0(x, y)$ and neighbor pixels are constituted by a Gaussian surface, so the mathematical expression is formula 1:

$$p = Aexp(-\frac{r^2}{B}) \tag{1}$$

In this formula, $r^2 = (x - x_0)^2 + (y - y_0)^2$, and (x_0, y_0) corresponds to a central location of Gaussian surface, A corresponds to the maximum value of the Gaussian surface, and correspondence with magnitude. The larger the magnitude, the greater the value of A; B corresponding to the spot size of the star, the smaller the size of the star, the smaller the value of B. The above equation with four unknowns. In order to obtain analytic equations, equation parameters such as x_0, y_0, A, B must to be get. However, the above equation is a nonlinear exponential function, analytic fitting parameters is very difficult. When taking the origin of the coordinates (x_0, y_0) into consideration, the above equation containing only A, B two parameters. Logarithm of both sides of the

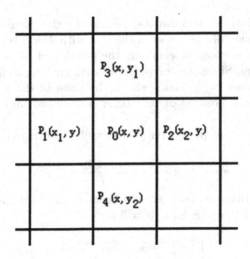

Fig. 3. Stellar location and adjacent gray distribution.

equation, there is formula 2:

$$\ln(p) = \ln(A) - \frac{r^2}{B} \tag{2}$$

Assuming:

$$y = \ln(p), x = r^2, a = -\frac{1}{B}, b = \ln(A) \tag{3}$$

There is:

$$y = ax + b \tag{4}$$

Obviously, the formula 4 is a linear function. It can be obtained coefficients a, b by a linear least-squares fitting method.

Assuming

$$S = \sum_{i=1}^{n}(y_i - y)^2 = \sum_{i=1}^{n}(y_i - ax_i - b)^2 \tag{5}$$

Taking the logarithm on both sides of formula 4 and setting logarithmic zero,

$$\begin{cases} \frac{\partial S}{\partial a} = 0 \\ \frac{\partial S}{\partial b} = 0 \end{cases} \tag{6}$$

It will be:

$$\begin{cases} a = \frac{k\sum y_i x_i - \sum y_i \sum x_i}{k\sum x_i^2 - (\sum x_i)^2} \\ b = \frac{\sum x_i^2 \sum y_i - \sum x_i \sum y_i x_i}{k\sum x_i^2 - (\sum x_i)^2} \end{cases} \tag{7}$$

In formula 7, k is the number of data. The coefficients, a and b, obtained from the above can help to restore the coefficients A and B in formula 2. Formula 7 shows that precision magnitude can be obtained by fitting Gaussian surface with a known star location coordinates. To simplify the calculations, the fitting position is the position of the center of the star with 4 or 8 adjacent pixels.

In the above calculation, determining the position of the stars becomes the key to the algorithm. Since the complex Gaussian surface, to facilitate the calculation, the Gaussian curve fitting should be done in the x and y direction respectively, then, the position of the stellar is obtained by finding the maximum value of the curve.

If we assume constant parameter y in formula 2, then in the x direction, it will be:

$$p = A exp(-\frac{(x - x_0)^2 + (y - y_0)^2}{B}) \tag{8}$$

Taking the logarithm on both sides of formula 8 and bringing $x1 = -1 x2 = 0 x3 = 1$ to it, solution of this formula will be:

$$\begin{cases} B = \frac{2}{2\ln(p_2) - \ln(p_1) - \ln(p_3)} \\ x_0 = \frac{B}{4}(\ln(p_3) - \ln(p_1)) \end{cases} \tag{9}$$

Obviously, x_0 is the star coordinates in the x direction, regardless of its size and y. Since the X-axis and Y-axis are symmetrical in the Gaussian surface relative to the coordinate origin, the star coordinate in y direction can be get in the same way, which is shown in formula 10.

$$\begin{cases} B = \frac{2}{2\ln(p_2) - \ln(p_4) - \ln(p_5)} \\ x_0 = \frac{B}{4}(\ln(p_5) - \ln(p_4)) \end{cases} \tag{10}$$

2.2 Fast Divisional Matching-Based Star Pattern Recognition Algorithm

In the basic star gallery, which stored a standard stellar parameters, their vector form is celestial coordinates which is represented by red latitude, in this paper, we use the basic celestial coordinates red latitude to identify. In the star sensor the images to be identified are two-dimensional gray-scale image, the two-dimensional X, Y coordinates with red latitude, and gray-scale image coordinates with magnitude. Usually in the identification, X, Y must be converted to red latitude. Unfortunately, due to before recognition, star sensor point is not completely sure, so that we can only get from a relative red latitude from X, Y coordinates, but can not get the absolute end, to this end, using the Hausdorff distance between the relative position of the star pattern and the satellite library as a criterion to conduct star identification.

Despite the lack of precise red longitude coordinates from the star sensor, however, depending on the structure and the relative position of the star field is kept substantially constant, set to be recognized star Pictured $A = \{a1, ..., ak, ..., ap\}$ star standard library $B = \{b1, ..., bj, ..., bq\}$, the improvement of the Hausdorff distance between them is defined as:

$$H = \sum_k min(d_k) \tag{11}$$

In this formula,

$$d_k = w_1(a_{1k} - (b_{1j} - b_{1i})) + w_2(a_{2k} - (b_{2j} - b2i)) + w_3 d_{smk} \qquad (12)$$

It represents the relative weighted distance between the k-th star, in A that to be identified, and the j-th star in star database B. W_i (i = 1,2,3) is the weight value. Usually, W_1 should equal to W_2, and i is the serial number of i-th star in star database, where j is the serial number of j-th star in star database. The third item represents the changes in magnitude. Due to the magnitude of the error is relatively large, the value of W_3 should be less than W_1. The distance between magnitude can be expressed as:

$$d_{smk} = \mid a_{3k} - b_{3j} \mid \qquad (13)$$

As the magnitude of the change has a great impact on the star pattern recognition, especially the star sensor threshold of exposure and weak star, due to the dynamic effects of noise and star sensor, when hidden, particularly large impact on the star pattern. Taking the impact of changes in magnitude into account, Eq. 12 can be rewritten as:

$$d_k = exp\left(\frac{d_{smk}}{d_0}\right)((a_{1k} - (b_{1j} - b_{1i})) + (a_{2k} - (b_{2j} - b_{2i}))) \qquad (14)$$

In formula 11 Hausdorff distance is the sum of the minim distance between each recognized star and star database in recognized stars, when the star to be identified in the star sensor match with the star in the star database, it has the least sum of the minimum distance. This avoids noise due to individual stars appear larger mismatch problem that may arise.

When there is uncertainty about the direction of the star sensor completely, it not sure star general area in the repository. Due to the number of stars in the all star database, more recognition speed is slow. Assuming N stars in star database, calculated by the formula 11, complexity of a complete minimum H distance calculating is N_2. If the standard star database is divided into M region, with N/M stars in each region, the complexity will reduce to N_2/M. Theoretically, there is M times faster than matching in whole star database, where the bigger M can get the faster calculation speed. However, the size of each area, should be greater than that of star sensor field, and each area should have enough redundancy for matching in a subregion.

Assuming that stars in the sky is evenly distributed, and taking the unit of length to be the radius of sky, the number of stars is $N/4\pi$. Assuming that the area size is $K \times K$, and the size of the FOV of star sensors is $L \times L$, one area includes the area of star sensor at least, meeting $K \times K \geq 2L \times 2$. Planar area is approximation for each region:

$$S_{partial} \approx K^2 \qquad (15)$$

An area which does not overlap region is:

$$S_{efficitive} \approx (K - L)^2 \qquad (16)$$

The rest is the redundant area. Assuming that each area is equal, the integral area number is:

$$NUM = \frac{S_{full}}{S_{efficitive}} \approx \frac{4\pi}{(K-L)^2} \tag{17}$$

Each number of stars in the area of approximation is:

$$M = S_{partial} \times \frac{N}{4\pi} \approx \frac{N}{4\pi}K^2 \tag{18}$$

The total number of calculations is:

$$Count = M^2 \times NUM \approx \frac{N^2 K^2}{4\pi(K-L)^2} \tag{19}$$

For the derivation and ordering derivative equaling to zero, it can get $K = 2$. When $K = 2$, the calculation number corresponds to the minimum value with $N^2 L^2/\pi$.

2.3   Divisional Strategies for Standard Star Database

By formula 19, in order to improve the speed of star pattern recognition, the standard star database should be divided into different regions. Because the star standard library is stored according to the spherical coordinates of latitude and longitude, the coverage of the latitude and longitude coordinates is not uniform to the certain star sensor in different latitude and longitude position, the star database segmentation is not uniform. Figure 4 shows the longitude and latitude area covered by star sensor.

The figure shows:

$$ab = 2ac\sin(\angle acb/2) \tag{20}$$

$$ac = oa\sin(\angle aoc) \tag{21}$$

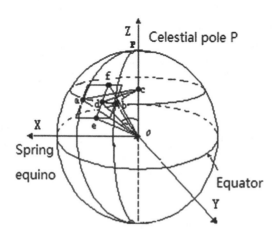

Fig. 4. Conventional diagram of star database segmentation.

So:

$$ab = 2 \times ac \times \sin(\angle acb/2) \times \sin(\angle aoc) \qquad (22)$$

$$ab = 2 \times oa \times \sin(\angle aob/2) \qquad (23)$$

By formulas 22 and 23 it can be obtained:

$$\sin(\angle aob/2) = \sin(\angle acb/2) \times \sin(\angle aoc) \qquad (24)$$

$\angle aob$ is the FOV of star sensor expressed as θ. $\angle acb$ is latitude expressed as α, and $\angle aoc$ is complementary angle corresponding longitude expressed as $90 - \delta$. The formula 24 will change to:

$$\sin(\theta/2) = \sin(\alpha/2) \times \cos(\delta) \qquad (25)$$

Along the longitude direction, $\angle eof$ corresponding to the amount of change is the longitude of $\triangle\alpha$, therefore, star database can be split directly along the longitude of the star sensor based on the field size.

A star sensor's FOV is 100×100, and each child area overlaps. The segment of star database calculated from formula 25 as shown in Table 1. The actual interval in the table is 2 times to division by $360°$.

Table 1. Division of star database.

No	Latitude range	Theoretical interval in degrees	The actual interval in degrees	Number of segments
1	$-90\sim-70$	360	360	1
2	$-80\sim-60$	90	180	4
3	$-70\sim-50$	30.5116	72	10
4	$-60\sim-40$	20.3220	48	15
5	$-50\sim-30$	15.6731	36	20
6	$-40\sim-20$	13.1018	30	24
7	$-30\sim-10$	11.5669	24	30
8	$-20\sim0$	10.6490	24	30
9	$-10\sim10$	10.1559	24	30
10	$0\sim20$	10.6490	24	30
11	$10\sim30$	11.5669	24	30
12	$20\sim40$	13.1018	30	24
13	$30\sim50$	15.6731	36	20
14	$40\sim60$	20.3220	48	15
15	$50\sim70$	30.5116	72	10
16	$60\sim80$	90	180	4
17	$70\sim90$	360	360	1
Total				298

3 Related Work

Since the first CCD-based star tracker was developed by Salomon in 1976 [7], great advancements in star identification have been made in about four decades. Many faster and more reliable methods were proposed from the 1990's [3]. Scholl proposed a method based on inter-star angles ordered by their relative brightness [8]. His method aimed at the search process acceleration with less time than the classical multi-step star identification method proposed by Baldini [9]. However, Scholl's method retains the $O(nf^2)$, so many faster techniques were proposed in the following years. To reduce the search time much further, a method using a "k-vector" to search the database in an amount of time independent of the size of the database [10] was proposed by Mortari. With this method, the search time for a single star-pair would be $O(k)$. Guangjun [11] proposed method based on feature extraction in 2007, using the inter-star angles and the angle made by two stars relative to a central star, which was similar to Liebe [12]. He uses a linear database search running in $O(n)$ time, while feature extraction time remains $O(f \lg b)$. In 2008, Kolomenkin [13] proposed a modification of the SLA algorithm [14] to reduce the time spent cross-checking the results of the k-vector. While the algorithm performs the cross check $O(k/f)$ faster than the SLA which take $O(k^2)$-time, it calculates $O(f^2)$ more inter-star angles, and k-vector searches, each of which takes $O(k)$-time, contributing an increase of $O(kf^2)$-time.

On the other hand, some non- dimensional algorithms and recursive star identification methods are proposed to improve the robustness of star identification. Rousseau computes the attitude for each star triangle with the sine of star-triangle interior angles, and the final analytic time of is $O(kf \lg f \lg n)$ [15]. Samaan reduced the recursive mode time was to speed the selection of stars for recursive star identification [16]. One of his methods uses the Mortari's Spherical Polygon-Search (SP-Search) [17,18], which uses a k-vector 3 times to find the stars within calculated x, y, and z ranges in inertial space. Each of the three database searches takes $O(k)$ time, while the cross-comparison takes $O(k^3)$-time. The another of his methods uses the Star Neighborhood Approach (SNA) which takes $O(b)$-time to find candidate stars, if b stars are identified. It is uncertain how many successive iterations would be necessary to ensure that all the stars in the given field of view have been found, other than it is most likely bounded by $O(fb)$.

4 Performance Evaluation

Using of statistical simulation, the extraction accuracy is analyzed by comparing Gaussian surface fitting with centroid method in [3].

Simulation Condition 1: The pixels position of theoretical centroid is $(4.7, 4.1)$, with $\sigma = 1$, and fitting spot size is 3×3 pixels. The result of 100 times on average is shown in Fig. 5.

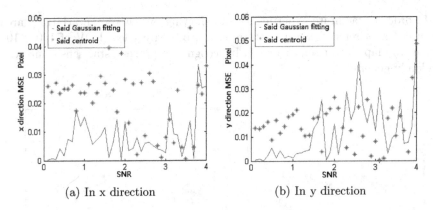

(a) In x direction (b) In y direction

Fig. 5. Different SNR results of two different methods in simulation 1

Simulation Condition 2: The pixels position of theoretical centroid is $(4.5, 4.5)$, with $\sigma = 1$, and fitting spot size is 3×3 pixels. The result of 1000 times on average is shown in Fig. 6.

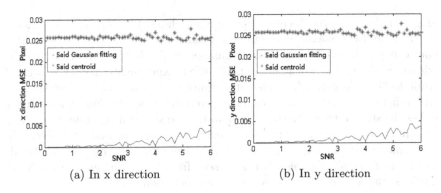

(a) In x direction (b) In y direction

Fig. 6. Different SNR results of two different methods in simulation 2.

The above results indicate that the window is calculated for participating quantized pixel size. When the centroid position is different, the distribution of image within the window is asymmetric, whereby the accuracy is different. When the SNR is relatively small, the centroid extracting accuracy is limited because of the image of an asymmetric distribution in the window. On the other hand, noise is mainly restricted in Gaussian fitting, so Gaussian fitting has higher accuracy than the centroid method. Therefore, for the establishment of high-precision imaging model, using Gaussian fitting method is a better choice.

Simulation Condition 3: Under different conditions, a simulation of fast divisional matching-based star pattern recognition algorithm was done in whole star database with the magnitude between 0 and 6.

In order to facilitate experiments and showing, we choose an arbitrary area $60° \times 60°$ as a reference star pattern, and arbitrarily select FOV of 100×100 regions to map to star sensor image. In recognition process, star sensor image is randomly generated.

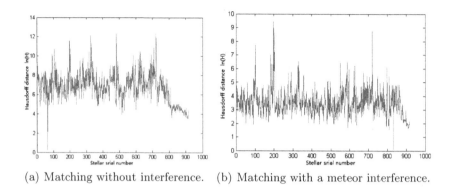

(a) Matching without interference.   (b) Matching with a meteor interference.

Fig. 7. Results of star pattern matching with random sensor image.

Figure 7 shows the results of star pattern matching. The y-axis represents the H distance, and the x-axis represents the serial number of stars. The matching results can be obtained by the serial number in the figure. According to the maximum of the star sensor possible noise, when experimenting, set the noise of latitude to 36 arc-second, the noise of magnitude to 0.5. It can be clearly seen from the figure that H distance is suddenly reduced when the star sensor image matches the star pattern. On the other hand, if the sensor image does not match the reference star pattern, H distance is relatively large and random. Figure 7(b) shows a matching result of a meteor interference with 1 arc-minute of latitude noise and 1 of magnitude noise. It can be seen from the figure, even under harsh conditions, the algorithm can identify the correct result.

5   Conclusion

Based on the analysis of star sensor imaging principle, this paper proposes a minimum use of star sensor relative space position Hausdorff distance map recognition method. This method is based on the spatial structure of stars similar principles, avoiding the complex feature extraction algorithm, but takes full advantage of all the information obtained by the star sensor. Experimental results show that the minimum matching method based on the Hausdorff distance, it is possible to obtain better recognition accuracy. For in the absence of prior knowledge, identify areas for full Star slower problem, the paper presents the basic methods star database partitions, partition star identification, so that the recognition speed is improved.

Acknowledgments. This research was supported in part by the NSF of China (Grant No.61272470, 61305087); the Fundamental Research Funds for the Central Universities, China University of Geosciences (Wuhan) (Grant No. CUGL120284, CUGL120289, CUG120114).

References

1. Ju, G.: Autonomous star sensing, pattern identification, and attitude determination for spacecraft: an analytical and experimental study (2001)
2. Eisenman, A.R., Liebe, C.C., Joergensen, J.L.: New generation of autonomous star trackers. In: Aerospace Remote Sensing 1997. International Society for Optics and Photonics, Conference Proceedings, pp. 524–535 (1997)
3. Spratling, B.B., Mortari, D.: A survey on star identification algorithms. Algorithms **2**(1), 93–107 (2009)
4. Pal, M., Bhat, M.S.: Star sensor based spacecraft angular rate estimation independent of attitude determination. In: 2013 IEEE International Conference on Control Applications (CCA), pp. 580–585. IEEE (2013)
5. Wang, H.-Y., Fei, Z.-H., Zhang, C.: An improved star pattern identification algorithm based on main star. Opt. Precis. Eng. **17**(1), 220–224 (2009)
6. Kandiyil, R.: Attitude determination software for a star sensor (2010)
7. Salomon, P., Goss, W.: A microprocessor-controlled ccd star tracker. AIAA paper, pp. 76–116 (1976)
8. Scholl, M.: Star-field identification for autonomous attitude determination. J. Guidance Control Dyn. **18**(1), 61–65 (1995)
9. Baldini, D., Barni, M., Foggi, A., Benelli, G., Mecocci, A.: A new star-constellation matching algorithm for satellite attitude determination. ESA J. **17**, 185–198 (1993)
10. Mortari, D., Neta, B.: K-vector range searching techniques (2014)
11. Zhang, G., Wei, X., Jiang, J.: Full-sky autonomous star identification based on radial and cyclic features of star pattern. Image Vis. Comput. **26**(7), 891–897 (2008)
12. Liebe, C.C.: Pattern recognition of star constellations for spacecraft applications. IEEE Aerosp. Electron. Syst. Mag. **8**(1), 31–39 (1993)
13. Kolomenkin, M., Pollak, S., Shimshoni, I., Lindenbaum, M.: Geometric voting algorithm for star trackers. IEEE Trans. Aerosp. Electron. Syst. **44**(2), 441–456 (2008)
14. Mortari, D.: A fast on-board autonomous attitude determination system based on a new star-id technique for a wide fov star tracker. Adv. Astronaut. Sci. **93**, 893–904 (1996)
15. Rousseau, G.L., Bostel, J., Mazari, B.: Star recognition algorithm for aps star tracker: oriented triangles. IEEE Aerosp. Electron. Syst. Mag. **20**(2), 27–31 (2005)
16. Samaan, M.A., Mortari, D., Junkins, J.L.: Recursive mode star identification algorithms. IEEE Trans. Aerosp. Electron. Syst. **41**(4), 1246–1254 (2005)
17. Parish, J.J., Parish, A.S., Swanzy, M., Woodbury, D., Mortari, D., Junkins, J.L.: Stellar positioning system (part i): applying ancient theory to a modern world. In: Astrodynamics Specialist Conference Proceedings
18. Woodbury, D., Parish, J.J., Parish, A.S., Swanzy, M., Mortari, D., Junkins, J.: Stellar positioning system (part ii): overcoming error during implementation. Astrodynamics Specialist Conference Proceedings

Performance Evaluation of HTTP and SPDY Over a DVB-RCS Satellite Link with Different BoD Schemes

Luca Caviglione[1(✉)], Alberto Gotta[2], A. Abdel Salam[3], Michele Luglio[3], Cesare Roseti[3], and F. Zampognaro[3]

[1] Institute of Intelligent Systems for Automation (ISSIA),
National Research Council of Italy (CNR), Genoa, Italy
`luca.caviglione@ge.issia.cnr.it`
[2] Information Science and Technology Institute (ISTI),
National Research Council of Italy (CNR), Pisa, Italy
`alberto.gotta@isti.cnr.it`
[3] Department of Electronics Engineering,
University of Rome "Tor Vergata", Rome, Italy
{`abdel.salam,roseti,zampognaro`}`@ing.uniroma2.it`, `luglio@uniroma2.it`

Abstract. The rapid evolution of the Web imposes the need of enhancing the HTTP over satellite channels. To this aim, SPDY is a protocol engineered to reduce download times of content rich pages, as well as for managing links characterized by large Round Trip Times (RTTs) and high packet losses. With such features, it could be an efficient solution to cope with performance degradations of HTTP over satellite. In this perspective, this paper compares the behaviors of HTTP and SPDY over a DVB-RCS satellite link. To conduct a thorough set of tests over a realistic scenario, we used the Satellite Network Emulation Platform (SNEP). In addition, we evaluated how different Bandwidth on Demand (BoD) methods impact over the retrieval of a page. Results clearly indicate that SPDY could be an effective solution to deliver Web contents over satellites in a more efficient manner.

Keywords: Networking protocol · Satellite network · BoD · SPDY · HTTP · DVB-RCS

1 Introduction

Nowadays, satellite communication is one of the preferred solutions for accessing the Internet while moving, and it is also the main choice to deploy connectivity in rural areas, or in developing Countries. Due to the physical characteristics of the link, especially long delays and high error rates, many protocols could experience performance degradations. For instance, mitigation of the impact of the high Round Trip Time (RTT) affecting GEO channels on TCP performance has been a prime research topic for years (see, e.g., [1] and references therein).

© ICST Institute for Computer Sciences, Social Informatics and Telecommunications Engineering 2016
I. Bisio (Ed.): PSATS 2014, LNICST 148, pp. 34–44, 2016.
DOI: 10.1007/978-3-319-47081-8_4

However, to effectively pursue the vision of the future Internet, satellites must also handle Web traffic, which is increasing both in terms of volumes and complexity [2]. In fact, modern web pages do not primarily rely on the *main* object containing the HTML code, but they also need several *in-line* objects. The evolution of the Web heavily requires in-line objects embedding interactive services, and content-rich graphic elements. As a consequence, the legacy *page-by-page* model should be updated, along with related protocols, such as the HTTP.

To partially fulfill such issues impairing the original HTTP, the HTTP/1.1 [3] introduced multiple connections to increase concurrency, and pipelining to send multiple object requests over a single TCP connection without waiting for a response. Even if such additions improve the performance over satellites, they are not definitive [4]. In fact, the server must generate responses ordered as the requests were received, thus limiting gains and possibly leading to blocking. Nevertheless, parallelism of HTTP/1.1 is usually limited (i.e., 7 connections in standard browsers), and not supported by-default by many servers.

On the contrary, SPDY is engineered to reduce download times of content-rich pages, as well as for managing wireless channels, which can be characterized by large RTTs and high packet losses [5]. Especially, to overcome to HTTP limitations, SPDY introduces:

- *multiplexed requests* - the number of concurrent requests that can be sent over a single connection is unbounded and properly handled by a framing layer;
- *prioritization* - retrievals of in-line objects composing a page can be properly scheduled, as to avoid congestion or to enhance the Quality of Experience (QoE). For instance, the client could fetch contents enabling to "understand" a page, even if incomplete;
- *header compression* - since the more sophisticated pages may need up to 100 requests, enforcing compression prevents bandwidth wastes due to duplicated headers;
- *server push* - contents can be pushed from servers to client without additional requests.

From an architectural point of view, the previous features are grouped within an high-level framing layer, which tunnels data into a single SSL/TCP connection. Hence, SPDY could be a suitable solution for the delivery of Web contents over satellite links. While the performance of HTTP has been extensively studied in literature both for wired [6] and satellite networks [7], a complete understanding of SPDY is still an open research problem. Moreover, many works focus on evaluating the protocol over wired and IEEE 802.11/cellular mobile scenarios [8]. For what concerns satellites, from our best knowledge, [9,10] are the only previous attempts.

Therefore, this paper compares HTTP and SPDY when used over a satellite link. To this aim, we used the Satellite Network Emulation Platform (SNEP) to conduct tests on a realistic DVB-RCS environment, with different Bandwidth on Demand (BoD) schemes. The contributions of this work are: (*i*) to understand the most relevant behaviors of SPDY when used over a realistic DVB-RCS channel; (*ii*) to provide a comparison between HTTP and SPDY emphasizing the impact of inline objects; *(iii)* to understand whether SPDY could be a suitable tool to enhance satellite communications in place of middleboxes.

The remainder of the paper is structured as follows: Sect. 2 describes the used testbed, while Sect. 3 discusses the measurement methodology. Section 4 presents the collected results, and finally, Sect. 5 concludes the paper.

2   Description of the Tesbed

To evaluate the performance of SPDY and HTTP over a satellite link we used the Satellite Network Emulation Platform (SNEP), developed by the University of Rome "Tor Vergata" [11]. Specifically, it can emulate different aspects of a DVB-RCS satellite access, for instance, the delay, the bitrate and TDMA framing, as well as BoD algorithms to dynamically assign the capacity on the return link. Figure 1 depicts the testbed implemented via the SNEP framework. Specifically, it is composed by:

- a gateway (GW) (or the access router);
- the network control centre (NCC) (or the hub);
- the satellite channel (SAT);
- the satellite terminals (ST);
- the user terminals (UTs), connected to the STs through a Local Area Network (LAN).

Each component used for the emulation is built via a Linux-based machine (version 2.6), and the needed functionalities are implemented through software modules running both in user and kernel space. To configure its behaviors (e.g., assign a fixed amount of bandwidth) a set of additional commands are made available through the Linux traffic controller `tc`. To manipulate traffic, ethernet frames are brought in the user space and then processed by an application-layer agent. The BoD portion is based on the DVB-RCS signaling, which is used to negotiate resources among different STs. Moreover, to emulate the DiffServ queuing discipline of DVB-RCS, packets are stored in a buffer implementing multiple parallel queues, which are served with different priorities.

For what concerns the hardware, both the Web client and the server are based on quad-core PCs with 32 GB RAM. To have a proper support for the SPDY protocol, we used Mozilla Firefox (24.0).

To capture data, we used the Wireshark sniffer with the SPDYShark extension enabling to decode protocol messages and to inspect its relevant parameters. When TLS/SSL encryption is used, we configured Wireshark to use the proper SSL keys to decrypt/decode the gathered data.

Table 1 reports the key configuration parameters characterizing our tests.

To emphasize the most critical behaviors when retrieving pages with different protocols, we created ad-hoc HTML pages for testing purposes. Specifically, to stress the iterations of the request-response exchange, each page contains a very large number of in-line objects, i.e., 640 small images. The main object implementing the hypertext (the `main.html`) has a size of 24.8 Kbytes. We point out that the test page has been crafted to highlight performance aspects of HTTP/SPDY when acting over high RTT links.

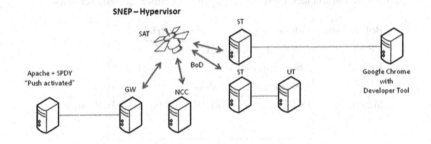

Fig. 1. Testbed used for evaluating SPDY and HTTP over a satellite network.

Table 1. TCP and Web server configurations.

TCP configuration	TCP Cubic (with optimized parameters for satellite [12])
	`tcp_moderate_rcvbuf` enabled
	$IW = 10$
Web server	Apache/2.2.22
	SPDY: Mod_SPDY 0.9.3.3
	Server Push (`X-Associated-Content header`)
	with different % for the pushed objects
	Apache KeepAlive settings enabled/disabled

3   Measurement Methodology

Planned measurements aim at comparing the behavior of SPDY with the most popular versions of the HTTP protocol, especially when different BoD mechanisms are deployed in the DVB-RCS return link. To this aim, a single user terminal (UT1) is connected to the virtual satellite terminal (ST1), which is implemented by SNEP. The available bandwidth is 4 Mbit/s and 1 Mbit/s, for the forward and the return link, respectively. The round-trip propagation delay is set equal to 520 ms, which is typical for a GEO satellite link. Besides, SNEP introduces an additional delay taking into account the MAC layer, i.e., the TDMA overhead and latencies due to the framing of the DVB-RCS. Tested BoD methods are:

- CRA (Constant Rate Assignment) - all the time slots are permanently assigned to the target station for the whole simulation duration;
- RBDC (Rate Based Dynamic Capacity) or VBDC (Volume Based Dynamic Capacity) - slots are dynamically assigned to the return link, depending on the traffic incoming to the satellite terminal: RBDC considers the ingress data rate, while VBDC uses the cumulative volume of queued data to compute capacity requests;
- mixed: a mix of the previous.

Table 2. Parameters used for emulating the different BoD schemes.

BoD scheme	Values (avg. on 50 repeated trials)
CRA	228 slots at a constant rate
RBDC	228 rate-based slots
VBDC	228 volume-based slots
Mixed	18 CRA slots, 105 RBDC slots, 105 VBDC slots

The values used for each BoD scheme are summarized in Table 2.

To have a fair assessment, we modified some parameters ruling the behavior of the browser, both for the case of HTTP and SPDY. In fact, default values are optimized for the wired Internet. In more details:

- **network.http.connection-retry-timeout:** defines the amount of time before considering a connection attempt aborted. Upon expired, the browser will open a parallel backup connection. Since this parameters is set to 250 ms by default, having an RTT of more than 520 ms would lead to unneeded TCP connections. Thus, it has been set to 0 in our trials (i.e., deactivated);
- **network.http.pipelining:** enables HTTP pipelining, i.e., the browser can send multiple GET without waiting for a server response. Pipelining has been enabled and set to 32 simultaneous requests, at the maximum;
- **network.http.speculative-parallel-limit:** normally, set to 6, it specifies the number of half-opened sockets to be kept for frequently used destinations (e.g., Google APIs). To avoid unpredictable behaviors of the browser, as well as to reduce overheads on the satellite link, we imposed this parameter to 0;
- **network.http.spdy.timeout:** defines the amount of time to wait after the page is considered received completely and the used TCP connection is closed (i.e., the RST/FIN). The default value is 180 s and is generally suitable to handle AJAX-based interactive contents. However, since our tests are aimed to verify performance during the page load, this value has been reduced to 5 s, as to not add noise to the measured times. We point out that this value is equal to the default keep-alive option of the Apache 2.2.2 used in our tests, as to guarantee a fair scenario.

The Web server has been configured to support SPDY and different versions of HTTP, i.e., HTTP1.0, HTTP1.1, SSL/HTTP1.0, and SSL/HTTP1.1. We underline that since SPDY uses SSL by default, this choice has been made to provide a more fair comparison. Trials have been performed with an instrumented Firefox browser, and each trial has been repeated 50 times.

4 Performance Evaluation

This section presents the outcome of the performance evaluation of the different version of HTTP and SPDY using an emulated DVB-RCS satellite access.

For the sake of conciseness, we present some results only using the BoD method defined as "mixed" (see Sect. 3). Nevertheless, in Sect. 4.4 we will offer a vis-à-vis comparison among the different BoD schemes.

Since it will be largely used in the rest of this section, we define the Page Loading Time (PLT) as [13]: $PLT = T_C^i - T_R$, where T_C^i is the time at which the last i-th inline object composing the page is completely received ($i = 642$ in our tests), and T_R is the time when the first GET for the index.html is performed.

4.1 Impact of the Header Compression

Native header compression is one of the most important design choice of SPDY, since it allows reducing the amount of transferred bytes. Figure 2 summarizes the amount of data transferred to retrieve the test page.

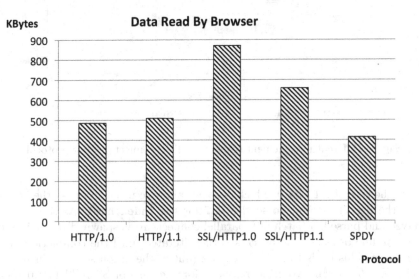

Fig. 2. Impact of the header compression per protocol.

Even if SPDY only needs 416 Kbytes to complete the transfer, such a result is very close to the cases when HTTP is adopted (~508 Kbytes). On the contrary, the real issue preventing the effectiveness of compression is due to SSL encryption, which accounts for overheads needed for the additional handshaking. In fact, the usage of SSL with the "traditional" HTTP leads to a significant increase of the transferred data: ~871 Kbytes for SSL/HTTP1.0 and ~659 Kbytes for SSL/HTTP1.1. Thus, the optimized design of SPDY in managing encryption [14] definitely plays a role.

4.2   TCP Connection Analysis

Understanding how TCP connections are managed by different protocols is mandatory to enhance their behaviors over satellite. Hence, Fig. 3 compares different evolution of the transport layer against the time.

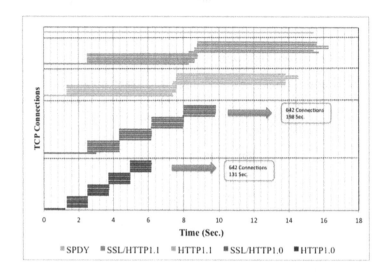

Fig. 3. Different management schemes of TCP connections per protocol.

For the case of HTTP1.0, which does not allow connection reuse, the browser opens the first TCP flow to send a GET for the index.html. As soon as it is received and parsed, a batch of 6 parallel connections is spawned to retrieve the needed in-line objects. The process is then iterated until the completion of the whole page. This leads to a PLT of 131 s (out of the graph scale), which is not acceptable. A similar evolution happens for the case of SSL/HTTP1.0 (again out of the graph scale). Yet, the need of performing additional exchanges for the SSL signaling, makes the first connection longer. Besides, encryption overheads account for longer transfer times, thus resulting into a PLT of 198 s.

When using HTTP/1.1, the first steps are still the same of the previous cases. Specifically, it uses a connection to retrieve the main object, and then uses the maximum allotted of 6 parallel TCP connections to retrieve additional contents. Then, it exploits the feature of connection reuse, and each one remains open to download 100 object (which is a limit imposed by the Apache web server). Recalling that our test page is composed by 642 objects, after hitting the limit of 600 items (i.e., 100 objects × 6 connections), a new batch of TCP connections is created. Such connections are mostly underutilized, since they are used to retrieve only a small fraction of data (equivalent to 42 objects only) (see Sect. 4.3). However, compared to HTTP1.0, the overall performance achieved in terms of PLT is better of an order of magnitude, i.e., PLT~10 s. The additional

5 s, for which the connections remain active are due to timeouts of Apache (i.e., the parameter **network.http.spdy.timeout**, as discussed in Sect. 3). Also SSL/HTTP1.1 behaves similarly, with times inflated by the overheads due to SSL (as explained earlier), accounting for an additional time of ~2 s in the PLT compared to the plain HTTP1.1.

Finally, in the SPDY case, the single-connection setup is clearly depicted. Thus, all objects are multiplexed into a unique TCP conversation. Once the page is completely received, the TCP connection is kept open by the browser for 5 s, as to maintain the same timeout period for an easier comparison. Its reduced overheads, and utilization of a unique (longer) connection, enables to better exploit the available bandwidth (even without using parallelization/pipelining). Then, SPDY has a PLT of 9 s, which is similar to HTTP1.1, but with a simpler complexity in the transport layer and supporting security. This is a plus, since satellite links are usually accessed through Performance Enhancement Proxies (PEPs) or middleboxes.

4.3   Throughput Analysis

Another important aspect to understand how the different Web protocols behave over a DVB-RCS link concerns the analysis of the throughput.

Figure 4 focuses on the HTTP1.0. Results indicate that the average rate is ~9 Kbyte/s. A deeper analysis reveals that the main cause is due to the usage of separate connections (one per object, 642 on the overall). Therefore, for each connection, a set-up and tear-down have to be performed, also worsened by the high values of the RTT, and impairments due to the slow-start. Similar considerations can be made when SSL is used.

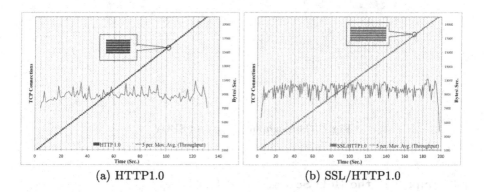

(a) HTTP1.0   (b) SSL/HTTP1.0

Fig. 4. Throughput analysis of HTTP1.0.

Figure 5 considers HTTP1.1. Since it implements pipelining and connection reusing, the latency impacts less on the behavior of the TCP. As a result, the achieved throughput is more than 200 Kbyte/s. Also in this case, SSL accounts for an overhead, slightly reducing the overall performances.

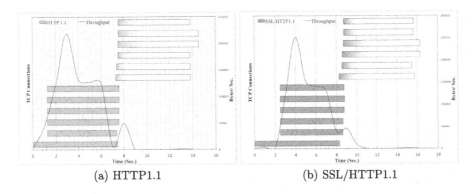

(a) HTTP1.1   (b) SSL/HTTP1.1

Fig. 5. Throughput analysis of HTTP1.1.

Finally, Fig. 6 shows the evolution of SPDY. Since, it uses a single connection, the delays introduced by the satellite network are absorbed (with the acceptation that they are experienced once, and not on a per-flow basis). Therefore, the achieved throughput is ~250 Kbyte/s, which is the best obtained value in our tests.

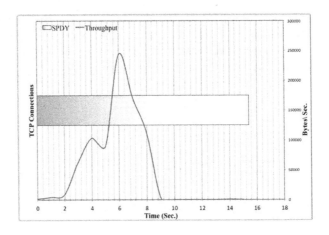

Fig. 6. Throughput analysis of SPDY.

4.4   Impact of the BoD Scheme

Figure 7 shows how the different BoD schemes impact on the PLT, for each protocol. Since previous results clearly show degradations due to the joint use of SSL and HTTP, the evaluation of the BoD scheme only considers the plain HTTP/HTTP1.1 and SPDY. In essence, the BoD increases the latency experienced by the application, which worsen the PLT. To highlight its impact, data transfers are performed on the return link.

As showcased, SPDY always outperforms the HTTP, and improvements increase for higher values of the RTTs, which characterize the VBDC and the RBDC schemes. In particular, SPDY is resilient enough to the increased latencies. In the worst case (i.e., the VDBC with an RTT of 1.6 s), its PLT is ~21 s, that is only 10 s greater than when using CRA. On the contrary, all the flavors of HTTP are greatly impaired by the VBDC, with a PLT ranging from ~150 s (for HTTP1.1) to ~300 s (for HTTP1.0).

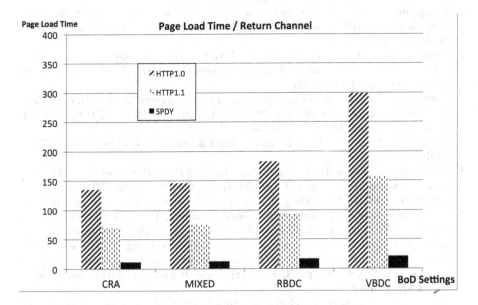

Fig. 7. PLT vs different BoD schemes.

5   Conclusions and Future Works

In this paper we compared the behaviors of different flavors of HTTP and SPDY over a DVB-RCS satellite link. To this aim, we used the Satellite Network Emulation Platform (SNEP) to conduct a thorough set of tests.

Results indicate that, owing to its single-connection architecture, SPDY is a promising solution to access the Web when using satellites. Also, it has a reduced TCP footprints, which can offload PEPs and middleboxes usually deployed in satellite service providers.

Future works aim at enriching the investigation, and also using real Internet Service Providers (ISPs) to better evaluate the feasibility of using SPDY as the unique technological enabler to bring modern Web contents via satellite platforms.

Acknowledgments. This work has been partially funded by the European Space Agency (ESA) within the framework of the Satellite Network of Experts (SatNex-III), CoO3, Task3, ESA Contract no. 23089/10/NL/CLP.

References

1. Sooriyabandara, M., Fairhurst, G.: Dynamics of TCP over BoD satellite networks. Int. J. Satell. Commun. Network. **21**(4–5), 427–449 (2003)
2. Caviglione, L.: Can satellites face trends? The case of Web 2.0. In: International Workshop on Satellite and Space Communications (IWSSC 2009), pp. 446–450, September 2009
3. Fielding, R., Gettys, J., Mogul, J., Frystyk, H., Masinter, L., Leach, P., Berners- Lee, T.: RFC 2616, Hypertext Transfer Protocol HTTP/1.1, Network Working Group, IETF (1999)
4. Nielsen, H.F., Gettys, J., Baird-Smith, A., Prudhommeaux, E., Lie, H.W., Lilley, C.: Network performance effects of HTTP/1.1, CSS1, and PNG. ACM SIGCOMM Comput. Commun. Rev. **27**, 155–166 (1997)
5. Belshe, M., Peon, R.: SPDY protocol - draft 3, Network Working Group, IETF (2012)
6. Heidemann, J., Obraczka, K., Touch, J.: Modeling the performance of HTTP over several transport protocols. IEEE/ACM Trans. Netw. **5**(5), 616–630 (1997)
7. Kruse, H., Allman, M., Griner, J., Tran, D.: Experimentation and modelling of HTTP over satellite channels. Int. J. Satell. Commun. **19**(1), 51–68 (2001)
8. Kim, H.J., Yi, G.S., Lim, H.N., Lee, J.C., Bae, B.S., Lee, S.W.: Performance analysis of SPDY protocol in wired and mobile networks. In: Jeong, Y.-S., Park, Y.-H., Hsu, C.-H.R., Park, J.J.J.H. (eds.) Ubiquitous Information Technologies and Applications. LNEE, vol. 280, pp. 199–206. Springer, Heidelberg (2014). doi:10.1007/978-3-642-41671-2_26
9. Cardaci, A., Caviglione, L., Gotta, A., Tonellotto, N.: Performance evaluation of SPDY over high latency satellite channels. In: Dhaou, R., Beylot, A.-L., Montpetit, M.-J., Lucani, D., Mucchi, L. (eds.) PSATS 2013. LNICST, vol. 123, pp. 123–134. Springer, Heidelberg (2013). doi:10.1007/978-3-319-02762-3_11
10. Cardaci, A., Celandroni, N., Ferro, E., Gotta, A., Davoli, F., Caviglione, L.: SPDY - a new paradigm in web technologies: performance evaluation on a satellite link. In: Proceedings of the 19th Ka and Broadband Communications Navigation and Earth Observation Conference, Florence, Italy, FGM Events LLC 239, October 2013
11. Luglio, M., Roseti, C., Zampognaro, F., Belli, F.: An emulation platform for IP-based satellite networks. In: IET Conference Proceedings, pp. 712–716 (2009)
12. Ha, S., Injong, R., Xu, L.: CUBIC: a new TCP-friendly high-speed TCP variant. ACM SIGOPS Oper. Syst. Rev. **42**(5), 64–74 (2008)
13. Wang, X.S., Balasubramanian, A., Krishnamurthy, A., Wetherall, D.: Demystifying page load performance with wprof. In: Feamster, N., Mogul, J. (eds.) Proceedings of the 10th USENIX conference on Networked Systems Design and Implementation (NSDI 2013), pp. 473–486. USENIX Association, Berkeley (2013)
14. Langley, A.: Transport Layer Security (TLS) Next Protocol Negotiation Extension, draft-agl-tls-nextprotoneg-00, Network Working Group, IETF, January 2010

Telecommunication System for Spacecraft Deorbiting Devices

Luca Simone Ronga, Simone Morosi$^{(\boxtimes)}$, Alessio Fanfani, and Enrico Del Re

University of Florence - CNIT, Via S. Marta 3, 50139 Florence, Italy
luca.ronga@cnit.it, simone.morosi@unifi.it

Abstract. In mission critical scenarios, fast and correct data reception is a crucial feature of telecommunication systems: particularly, to cope with unknown and fast variable channel state condition, with large Doppler shifts and low available power, incoherent and computationally light modulation techniques have to be considered. This paper deals with the design of suitable systems for a specific application fields, namely control operation of a satellite while it is placed into its orbit and the disposal of a satellite at the end of its life or the deep-space missions; moreover, a digital implementation of a receiver, based on DD-PSK modulation, is introduced which is perfectly compliant with these requirements.

Keywords: DD-PSK · Decommissioning · Deorbiting · Digital receiver

1 Introduction

In the next years, an easier space access by means of new low-cost satellite solutions such as micro or nanosatellites, could drive technology advances in communications. End-users and also satellite operators will take advantage of these opportunities. The grow of innovative satellite technologies will increase the global offer of broadcast telecommunication services [1].

The most interesting orbits for space based activities, namely GEO and LEO ones, are getting crowded by an huge number of space debris. Space debris are man-made objects like as dead satellites, upper stages, pieces from fragmentation or collision of parts and so on. Today the number of that space debris fragments is higher than some hundred million, as depicted in Table 1, and is growing exponentially [2]. Consequently, space debris may start colliding into each other and thus increasing the already high number of debris, towards a sort of collision reaction, known also as Kessler Syndrome that may lead to a space saturation [3].

Recent debris mitigation guidelines suggest to dispose and passivate a satellite at the end of its life [4,5]. This goal requires the use of a reliable communication link also when the satellite has limited functionalities, with a random and uncontrollable attitude or when the satellite could be defunct. In this scenario, the channel conditions may change rapidly, e.g., signal's amplitude could drop down or go up according to satellite tumbling rate. From a more general point of view, typical satellite channel impairments like heavy Doppler effect, phase

© ICST Institute for Computer Sciences, Social Informatics and Telecommunications Engineering 2016
I. Bisio (Ed.): PSATS 2014, LNICST 148, pp. 45–57, 2016.
DOI: 10.1007/978-3-319-47081-8_5

Table 1. Estimated orbital population

Size	Number	Mass %
>10 cm	>30,000	99.93
1 − 10 cm	>750,000	0.035
<1 cm	>166,000,000	0.035
Total	>166,000,000	6, 000 tonnes

distortion, harshly impair signal propagation: in these contexts traditional signal acquisition and synchronization techniques could end up being slow and inappropriate. In this paper a robust transmission technique is proposed which is based on the DD-PSK modulation [6]: this incoherent solution results to be extremely suitable for this context. Another positive feature of DD-PSK is the simple receiver's architecture: the absence of a PLL or of a carrier acquisition's circuits limits the system complexity.

This study is the result of a cooperation between University of Florence and D-Orbit S.r.l., an Italian company that develops a smart satellite device that decommissions satellites at the end of their operative life.

The paper is organized as follows: Sect. 2 provides an overview of the application's scenario and an estimation of the link budget. Section 3 proposes a complete description of the DD-PSK modem architecture and an analytical analysis of its operative principles. Section 4 discusses the performance of the whole transceiver architecture and shows the simulated results which have been obtained. Finally, conclusive remarks are given in Sect. 5.

2   Scenario

This section describes a possible scenario for a satellite decommissioning application. The goal is the design of a Telemetry and Tele-Command link (TM/TC) for a satellite placed in a polar Low Earth Orbit. The operative frequency is in S-Band: in particular ITU radio regulamentation and satellite standard [7,8] reserve the following sub-range:

– Frequency range: 2025 − 2110 MHz for Earth to space link;
– Frequency range: 2200 − 2290 MHz for space to Earth link.

The choose of S-band frequency was validated by a tradeoff analysis that has excluded the possible use of higher operative frequency, such as the X-Band (8 to 12 GHz). The major benefit of the use an X-band frequency is the considerably smaller physical dimension for the antenna that simplifies the integration on the satellite. On the other hand, the use of higher frequency brings stronger attenuations: the pathloss is increased by 14 dB and the link is greatly affected by other impairments such as ionospheric effects, scintillation, rain and fog attenuation [9]. Moreover, frequency value influences the performance of a ground station

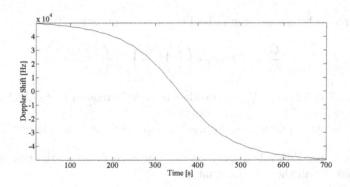

Fig. 1. Doppler shift S-curve

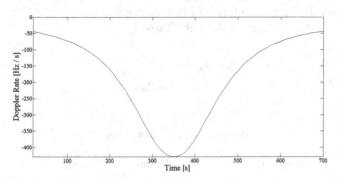

Fig. 2. Shift Doppler time variation

antenna: with an X-band frequency the HPBW angle becomes smaller and this consequently increases link's pointing losses; finally, the higher the frequency link, the higher the Doppler shift.

In LEO orbit the main channel impairment is due to the large and time-variant Doppler shift within a visibility window of satellite. The frequency shift can be represented by the S-curve, which is shown in Fig. 1: the maximum shift is within ±50 kHz. Another important figure is the Doppler shift rate that represents the time-variation of shift. The curve in Fig. 2 describes this feature: the worst case is achieved in the center of visibility window, when the satellite is closer to the ground station, and the Doppler rate is lower than −400 Hz per second.

The previous figures are compliant with CCSDS raccomandations, [8], that suggest a frequency acquisition range of at least ±150 kHz at 2 GHz and a minimum frequency sweep rate of 100 Hz/s.

The satellite is equipped with an hemispherical patch antenna dedicated to TM/TC link with antenna boresight allined with yaw satellite's axis. This configuration provides a stable signal only when satellite is in cruise mode, with a working attitude control system. With these hypothesis, the received power is easily computed by means of standard link budget equation [9–11].

$$\frac{C}{N_0} = EIRP_{TX}\left(\frac{1}{L}\right)\left(\frac{G}{T}\right)_{RX}\left(\frac{1}{k}\right)$$ (1)

Equation 1 define the C/N_0 ratio that is directly related to E_b/N_0 by means of formula in Eq. 2.

As for the other parameters in previous equation, we can assume that:

- $EIRP_{TX}$ is the Equivalent Isotropically Radiated Power of the transmitter;
- L summarizes all the link attenuations;
- k is the Boltzmann constant;
- the receiving station's figure of merite G/T is function of antenna noise temperature T_A and noise figure of the receiver;
- $R_b = 1/T_b$ is the bit rate of the link (Table 2).

$$\frac{E_b}{N_0} = \frac{C}{N_0}\frac{1}{R_b}.$$ (2)

Table 2. Link budget table

Parameters	Uplink	Downlink
	Worst Case	Worst Case
Satellite Slant Range [km]	2500	2500
Link Frequency [MHz]	2025	2200
Rate [kbps]	128	128
Transmitter Power [dB]	13.01	3.01
TX Antenna Gain [dB]	35	6
PathLoss [dB]	166.5	167.2
Polarization Loss [dB]	3	3
De-Pointing Loss [dB]	2.7	2.7
Atmospheric Loss [dB]	0.3	0.3
Scintillation Loss [dB]	10	10
RX Antenna Gain [dB]	6	35
Interconnection Loss [dB]	2.8	2.8
Noise Temperature [dBK]	27.8	13.7
EIRP [dB]	48.01	9.01
Figure of Merit [dB/K]	−21.8	21.3
Received Power [dBm]	−102.3	−112
C/N_0 [dB]	74.9	65.25
E_b/N_0 [dB]	23.9	14.18

In Fig. 3(a) and (b) the received power and the signal to noise ratio E_b/N_0 are represented in a visibility windows. These graphics are associated to a satellite with a perfectly working attitude control system. Such condiction is not always verified. For example, before the attitude stabilization or at the end of mission due to a fail of attitude control system, the satellite could casually lose stability. In this scenario, the channel results to be very unstable and the signal strength would follow the antenna radiation pattern dropping out completely when the antenna rotates away from ground station [12]. Figure 4(a) and (b) report the values of the previously considered curves as given by the link budget curves in a possible critical scenario. A high tumble rate has been supposed around all satellite's axes. The presumed angular rotation velocities are 20 deg/sec around Pitch axis, 10 deg/sec around Roll axis and 5 deg/sec around Yaw axis. A sequence of impulses highlights the signal instability and an adequate signal power is obtained only in short burst intervals whose duration is often less than 1 s. In this condition, a frequency acquisition becomes very hard due to quick amplitude fluctuations so that incoherent demodulation techniques could be the only feasible solution.

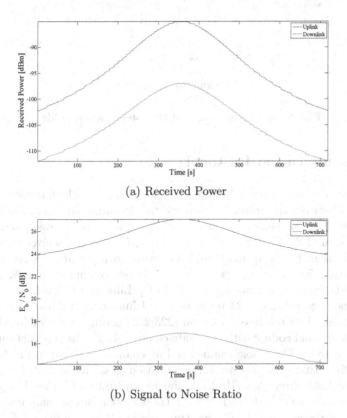

(a) Received Power

(b) Signal to Noise Ratio

Fig. 3. Link budget curve in stable attitude condition

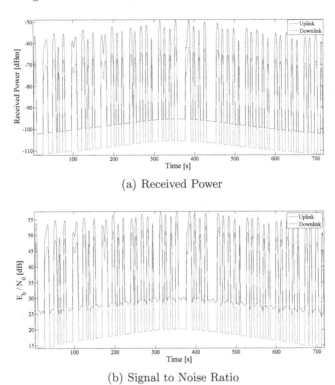

(a) Received Power

(b) Signal to Noise Ratio

Fig. 4. Link budget curve at the satellite's end of life

3   Communication Protocol

This section describes a possible communication protocol which is used to establish the link between the ground station and the decommissioning device installed on satellite. The transmission is a typical half-duplex communication, coming from the Ground Station that is the *Master Node*. The transceiver on satellite, *Slave Node*, is in receiving mode and it's ready to reply at each request from ground station. The approach is similar to a burst communication: Master and Slave node transmit short messages both in Up-Link and Down-Link.

Each message includes 223 bytes of useful information data. This message is then processed with a Reed Solomon (223,255) coding, a scrambling function and a convolutional coding with code rate equal to $\frac{1}{2}$. The final packet dimension becomes 530 byte. The whole channel coding chain is depicted in Fig. 5.

Supposing that a complete communication session include 10 messages exchange in both directions (Up-Link and Down-Link) and a session period is 1 s, as previously described, the minimum transmission rate is computed equal to 81600 bit per second. So, the desired transmission Rate has been defined equal to $2^{17} = 131072$ b/s.

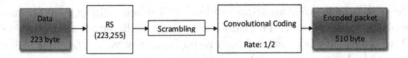

Fig. 5. Channel coding

4   Transceiver Architecture

The proposed transceiver architecture encompasses an analog front-end with a duplexer to divide receive and transmission chains, an IF section and a digital section implemented on a FPGA. A high level functional block of the transceiver is depicted in Fig. 6

The main functionalities which are implemented in the digital section are the signal modulation/demodulation, the IF to baseband frequency convertion and the frame coding/decoding.

The selected modulation is the Double Differential PSK (DD-PSK). Thanks to the double differential step, it's possible to demodulate the IF signal without a frequency acquisition or a phase estimation, even with a heavy Doppler shift [13]. This incoherent demodulator facilitates the implementation, increases the receiver's reliability and reduces the power consumption [6].

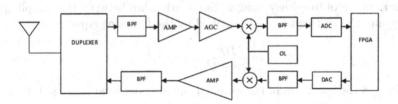

Fig. 6. Transceiver block diagram

4.1   Transmitter

The transmitted signal is PSK modulated; its information messages dictates the phase difference between three consecutive bits. The block representation of the DD-PSK modem is depicted in Fig. 7.

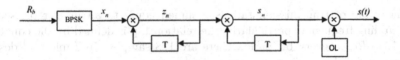

Fig. 7. DD-PSK modulator

The first B-PSK stage associates to bit "1" a phase value $\varphi_n = 0$ and to bit "0" a phase value $\varphi_n = \pi$. Thereafter, the double differential step creates the DD-PSK signal that is converted to a radio frequency value, f_{OL} into S-band. The delay blocks introduce a delay equal to the symbol period $T = 1/R_b$.

The analytical form of the signal is:

$$x_n = e^{i\varphi_n} \tag{3}$$

$$z_n = x_n x_{n-1} = e^{i\varphi_n} e^{i\varphi_{n-1}} = e^{i\Delta\varphi_n} \tag{4}$$

$$s_n = e^{i\Delta\varphi_n} e^{i\Delta\varphi_{n-1}} = e^{i\Delta^2\varphi_n} \tag{5}$$

$$s(t) = A\cos(2\pi f_{OL}t + \Delta^2\varphi_n) \tag{6}$$

where:

φ_n is the phase of bit n;

$\Delta\varphi_n = \varphi_n - \varphi_{n-1}$ is the first order difference;

$\Delta^2\varphi_n = \Delta\varphi_n - \Delta\varphi_{n-1} = \varphi_n + 2\varphi_{n-1} + \varphi_{n-2}$ is the second order difference.

4.2   Receiver

The receiver implementation works at Intermediate Frequency IF and uses a delayed reply of the same signal to down-convert it to base band.

An important design parameter is the sampling frequency. In order to avoid the overlapping of frequency images, the relationship between the sampling frequency and the IF frequency is described by the following equation:

$$f_c = \frac{4f_{IF}}{2n+1}, \quad n = 0, 1, 2, .. \tag{7}$$

In Fig. 8 the spectrum of the sampled signal is depicted for $n = 0$: no signal overlapping is shown.

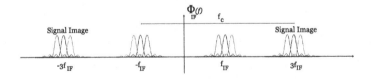

Fig. 8. Spectrum of the sampled signal

The value of transmission rate is assumed equal to an integer divisor of the sampling frequency: particularly, the relationship is defined by the equation $f_c = 128 \cdot R_b$; in every bit period there are 128 samples. In Table 3 all design parameters for the receiver are summarized.

The core of the receiver is the DD-PSK demodulator characterized by the two differential steps, which are depicted in Fig. 9. The main blocks of the demodulator are:

Table 3. Receiver design parameters

Transmission rate	$2^{17} = 131072\,\text{b/s}$
Intermediate frequency	$2^{22} = 4,194\,\text{MHz}$
Sample frequency	$2^{24} = 16,777\,\text{MHz}$
Maximum shift Doppler	$150\,\text{KHz}$

– the delay blocks T_b;
– the *integrate and dump*, activated by the synchronization circuit;
– the threshold comparator for the bit detection.

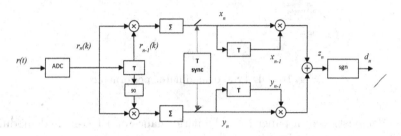

Fig. 9. Diagram of DD-PSK demodulator

The input signal, without noise contribute, is defined by the following equation:

$$r(t) = A\cos(2\pi((f_{IF} + \Delta f_{D_n})(t - \tau_{ch_n})) + \theta_n + \varphi_n) \qquad (8)$$

where:

φ_n is the phase of bit n;

f_{IF} is the Intermediate Frequency;

Δf_{D_n} is the shift Doppler on bit n;

τ_{ch_n} is the channel delay on bit n;

θ_n is the phase error on bit n.

By means of several analytical computations, the final receiver output signal is:

$$d_n = sgn[\frac{T_b^2 A^4}{4}\cos\Delta^2\varphi_n] \qquad (9)$$

Equation clearly shows that the result is independent from the Doppler Shift and/or the channel phase error: particularly, the output depends only on the two previous bits and their phase difference $\Delta^2\varphi_n$.

5   System Performance

A demonstration of DD-PSK robustness against the Doppler effect is obtained through a complete simulation of the receiver. The simulation has been performed in Simulink and the results are reported in Fig. 10. The bit error rate curves are nearly constant for different Doppler conditions. The results have been obtained using a random binary transmitted sequence which is formed by 10^7 bits.

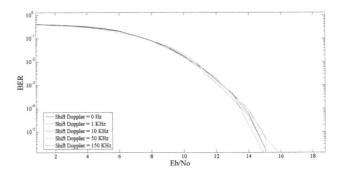

Fig. 10. DD-PSK demodulator performance

As demonstrated in paper [13], the performance of a DD-PSK demodulator can be computed as in Eq. 10.

$$P_{e_{DD-PSK}} = \exp[-\frac{E_b}{2N_0}(1 + \frac{N_0}{2E_b})^{-1}]$$ (10)

The DD-PSK Bit Error Probability is plotted in Fig. 11 where it is compared with BER curves of other binary modulations. Figure 11 shows that DD-PSK performance, in terms of Bit Error Rate, are about 4 dB lower than B-PSK coherent demodulation. This weakness could be exceeded introducing efficient channel coding techniques.

Finally, it's important to observe that the theoretical BER curves in Fig. 11 correspond with simulated results depicted in Fig. 10.

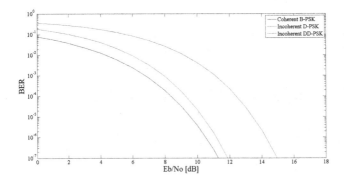

Fig. 11. Bit Error Rate for different modulation

5.1 Time Synchronization

In this section the synchronization issues are evaluated: they are deemed as a key element in the receiver design.

In the previous sections, the demodulator derivation is performed by assuming perfect synchronization. A key aspect of the DD-PSK receiver can be identified in the correct demodulation which can be performed with no signal pre-acquisition. Thanks to this peculiar feature this receiver can be proven robust in scenario with quick channel variations with no need to resort to long preamble sequences. An effective synchronization circuit will help to preserve this property and minimize acquisition time. The adopted frame of 512 bits is composed by a short preamble (16 bit), followed by the payload. When the preamble is acquired, the synchronization circuit generates a brief pulse that reset the *Integrated and Dump* in the demodulator. An example of this signal is represented in Fig. 12: the impulse period is nearly equal to 4.5 µs that is the frame period.

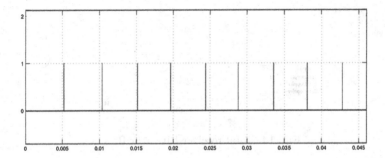

Fig. 12. Synchronization signal generated by Synchronization circuit

The performance of synchronization circuit is analyzed in Fig. 13. These curves show the frame loss probability, that is a missed preamble acquisition, in function of signal to noise E_b/N_0 and Doppler shift. We can conclude that the proposed synchronization circuit is insensitive to Doppler shift up to 50 kHz and it shows a satisfactory performance when E_b/N_0 is greater than 9 dB.

5.2 Overall Performance

The results of the whole receiver are shown in the following. To this aim we report the Bit Error Rate curves which are depicted in Fig. 14. The obtained BER results are similar to the curves in Fig. 10, i.e., the ones which are associated to an ideal DD-PSK demodulator. This result demonstrates the effectiveness of the global receiver architecture.

A further improvement of receiver performance is obtained by introducing channel signal coding [14]. An explanatory example of the improvement is given by BER curve in Fig. 15. In these simulation, the frame has been encoded with

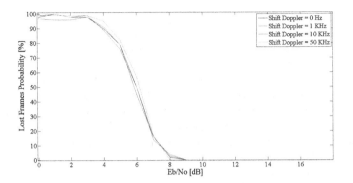

Fig. 13. Lost frame probability

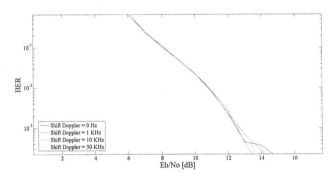

Fig. 14. Total receiver Bit Error Rate

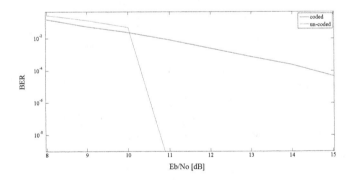

Fig. 15. Bit Error Rate with coded and uncoded frame

a Reed-Solomon (255,223) and a convolutional coding with coding rate equal to $1/2$. The performance increase is significant: with a signal to noise ratio, E_b/N_0, greater than $10\,\text{dB}$, the final Bit Error Rate is less than 10^{-8}.

6   Conclusion

This paper analyzes the possible channel conditions in mission critical scenarios and proposes a modem architecture for a reliable earth-space telemetry communication link which is based on the DD-PSK modulation and is able to meet the system requirements. The proposed solution has been validated through computer simulations considering all the potential impairments in the chosen scenario or in a deorbiting mission: the incoherent feature and the robustness to Doppler of the DD-PSK simplify the receiver architecture and afford a reliable link also in emergency scenario.

References

1. Morosi, S., Jayousi, S., Del Re, E.: Cooperative delay diversity in hybrid satellite/terrestrial DVB-SH system. In: Proceedings of 2010 IEEE International Conference on Communications, ICC 2010, Cape Town, South Africa (2010)
2. Reflections on Orbital Debris Mitigation Measures Prof. Richard Crowther. Chief Engineer, UK Space Agency
3. Scientist: Space weapons pose debris threat CNN, 03 May 2002. Articles.cnn.com. Accessed 17 Mar 2011
4. IADC Space Debris Mitigation Guidelines, 25 October 2002
5. European Code of Conduct for Space Debris Mitigation, 28 June 2004. Issue 1.0
6. Yuce, M.R., Liu, W., Damiano, J., Bharath, B., Franzon, P.D., Dogan, N.S.: SOI CMOS implementation of a multirate PSK demodulator for space communications. IEEE Trans. Circ. Syst. I Regul. Pap. **54**(2), 420–431 (2007)
7. ECSS-E-ST-50-05C Rev. 2: Radio Frequency and Modulation, ESA-ESTEC, 4 October 2011
8. CCSDS 401.0-B: Radio Frequency and modulation system, January 2013
9. Ippolito, L.J.: Satellite Communications Systems Engineering: Atmospheric Effects, Satellite Link Design and System Performance. Wiley (2008)
10. Maral, G., Bousquet, M.: Satellite Communications Systems: Systems, Techniques and Technology. Pearson Education, India (2009)
11. Wertz, J.R., Larson, W.J.: Space Mission Analysis and Design. Microcosm Press (1999)
12. Bruzzi, J.R., Jensen, J.R., Fielhauer, K.B., Royster, D.W., Srinivasan, D.K.: Telemetry recovery and uplink commanding of a spacecraft prior to three-axis attitude stabilization. In: 2006 IEEE Aerospace Conference (2006)
13. Ma, C., Wang, D.: The performance of DDPSK over LEO mobile satellite channels. In: Proceedings of the 2000 Asia-Pacific Microwave Conference (2000)
14. Shu, L., Costello, D.J.: Error Control Coding. Pearson Education, India (2005)

Quality of Service and Message Aggregation in Delay-Tolerant Sensor Internetworks

Edward J. Birrane III[(⊠)]

Space Department, Johns Hopkins University Applied Physics Laboratory,
Laurel, MD, USA
Edward.Birrane@jhuapl.edu

Abstract. We present traffic-shaping and message-aggregation algorithms that provide reservation-based quality-of-service mechanisms for delay-tolerant internetworks utilizing graph-based routing protocols. We define a Traffic Shaping with Contacts (TSC) method that alters the edge weights in a graph structure to represent service level specifications, rather than physical capacity. This adjustment allows existing routing mechanisms to implement bandwidth reservations without additional processing at the node. We define a Payload Aggregation and Fragmentation (PAF) algorithm that calculates preferred payload sizes over traffic-shaping contacts. PAF aggregates too-small payloads together and fragments too-large payloads to optimize contact capacities. Unlike other mechanisms, TSC/PAF are unaffected by heterogeneous physical, data-link, and transport layer protocols across an internetwork and require only minor modifications to internetwork-layer graph-routing frameworks. Simulation results show that together TSC/PAF reduce the number of messages in a sensor internetwork by 43 % while increasing the goodput of the network by 63 % over standard graph-routing techniques.

Keywords: Delay-tolerant networking · Congestion modeling · Traffic prediction · Quality of service · Fragmentation · Aggregation

1 Introduction

Delay-tolerant networks (DTNs) [1] operate in environments characterized by significant propagation delay, frequent link disruption, and a potentially heterogeneous set of communications hardware. These networks were conceived to enable packetized data communications across interplanetary distances and amongst multiple, heterogeneous spacecraft [2]. DTN architectures also apply to terrestrial internetworks, especially those that include space-based nodes. In terrestrial internetworks, delays and disruptions manifest not only from physical limitations but also from administrative, security, and service policies at local-network boundaries.

Emerging networking architectures, such as delay-tolerant sensor internetworks, provide periodic but deterministically available links. Such architectures use land/space/aerial vehicles to collect data from geographically separated sensor networks (or associated cluster heads). In this paper, we refer to such network architectures as *Challenged Sensor Internetworks* (CSIs) to distinguish them from opportunistically-routed DTNs and other Mobile Ad-Hoc Networks (MANETs).

© ICST Institute for Computer Sciences, Social Informatics and Telecommunications Engineering 2016
I. Bisio (Ed.): PSATS 2014, LNICST 148, pp. 58–75, 2016.
DOI: 10.1007/978-3-319-47081-8_6

The periodic and deterministic nature of CSIs allows them to pre-configure future contact opportunities to assist routing mechanisms. A structure for representing contact opportunities between nodes over time is a weighted, directed *Contact Graph* (CG). Given such a graph, paths are computed via any one of a number of graph-theoretic approaches. Since the contact description stored in a CG is generalized, networking functions that restrict their input to the CG operate independent of specific physical, data-link, and transport layer configurations. This is a required feature when considering end-to-end message exchange in CSIs that federate local networks of differing media access, protocols, or administrative privileges.

We investigate a delay-tolerant quality of service (DTQoS) approach for CSIs based on service-level agreements specifying data rates during times of active link connections. These agreements throttle internetwork traffic over individual constituent local networks. Quality of service, in this context, prevents capacity contention between internetwork traffic and pre-existing local network traffic. Our approach specifies how much active link time a constituent local network will devote to the internetwork. We further propose a proactive fragmentation/aggregation scheme to more efficiently bundle user data thus avoiding lower-level protocol fragmentation strategies and their inherent inefficiency.

We define a *Traffic-Shaping Contacts* (TSC) method and a *Payload Aggregation and Fragmentation* (PAF) algorithm to enforce DTQoS agreements and increase the effective use of contact residual capacity, respectively. Together, TSC and PAF provide four benefits in a CSI: (1) payload fragmentation puts small capacities in contacts to useful work, (2) calculation of payload size reduces the chance of downstream fragmentation, (3) aggregation of small data sets reduces messaging overhead, and (4) traffic shaping allows multiple user classes in the internetwork. These benefits can be realized with very little modification to existing CG-based routing schemes.

We analyze the performance of TSC and PAF in an exemplar sensor internetwork simulated using the ns-3 network simulator. Simulation results show these approaches increase network goodput, prevent starvation in networks with multiple classes of data, and allow local networks to enforce the amount of internetwork traffic they are required to carry.

The paper is organized as follows. Section 2 discusses the motivation for this work. Section 3 summarizes related work. Sections 4 and 5 provide overviews and analyses of the TSC and PAF algorithms, respectively. Section 6 presents an ns-3 simulation of these algorithms, demonstrates their performance in reference scenarios, and discusses results. Section 7 concludes the paper.

2 A CSI Architecture

A typical sensor internetwork architecture federates pre-existing, heterogeneous local networks to cover vast or remote regions in a cost-effective manner. We have previously identified three tiers within this architecture: data-generating networks (Tier 1), regionally mobile networks (Tier 2), and exfiltration networks (Tier 3) [3]. The networks in each tier may be separated by distance, by media access, by protocol support,

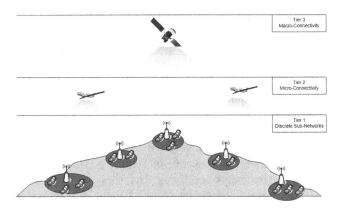

Fig. 1. A federated space-terrestrial sensor web provides applications with link capacity as a function of contact time [3].

or by policy. This architecture is illustrated in Fig. 1 where discrete sensing networks provide data to aerial data mules that forward sensed data to a satellite exfiltration layer.

Constituent networks in the CSI may have pre-existing policies, qualities of service, and classes of users. End-to-end data exchange across the internetwork must work within these existing policies, even though these policies differ in each local network.

Consider the example of three low-powered sensor networks (S1, S2, S3), each with a priority scheme (**H**igh, **M**edium, **L**ow) that dictates how sensed data are communicated from a cluster head to a mobile data mule (such as a jeep driving through each sensor network or an unmanned aerial vehicle) which passes through the three networks in the order S1, S2, S3. The data mule communicates collected data to a Low Earth Orbit (LEO) satellite during a pre-planned pass, with the observation that the short duration of the satellite pass is not sufficient to exchange all data collected. The nominal transmit queue for the data mule is shown in Fig. 2, where received data are kept in a first-come, first-served order. The data mule must have some way to prioritize the data from S1, S2, and S3, but each of these local networks has a different prioritization scheme. If the data mule queues data on a first-come, first-served basis, then the last sensor network reporting data (S3) is given no opportunity to commu-nicate in the upcoming satellite contact.

Figure 2 illustrates two problems that must be solved to negotiate service quality in an internetwork. First, some algorithm must order data in the transmission queue while balancing different local network service quality schemes. Treating the transmission

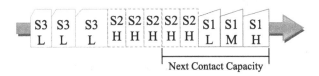

Fig. 2. The transmit queue of an internetworking mobile node must understand individual sub-network priorities to avoid starvation.

queue as First-In, First-Out (FIFO) is clearly unfair as in our example no messages from S3 will be communicated during the next contact. However, attempting to build a sorted transmission queue assumes that some comparison function can be built to consider data from amongst all local networks and produce a linear sorting of trans-missions over time. This is not the case when the priority of nodes is dependent upon system time, message age, messages delivered, or any other dynamic quantity. For this reason, classical schemes for packet-based service differentiation do not properly address this issue. Furthermore, even if such a function existed it would grow expo-nentially as local networks were added to the internetwork.

Second, reducing data volume reduces the loss associated with service differenti-ation. In Fig. 2, the more data in the transmit queue that can be communicated in the upcoming contact, the fewer data are queued for transmission later. As a sensing internetwork, payloads often share a common destination and therefore there is no advantage to repeating identical header information. The payloads of multiple small messages can be aggregated into a single larger payload to reduce overhead. Con-versely, a single large payload may be fragmented into multiple smaller payloads utilizing residual capacities from non-saturated contacts in the CG. Unused portions of the contact may, in aggregate, account for a significant amount of wasted transmission opportunities.

This paper is motivated by the need to (1) practically sort the transmission queue of a mobile node fed by multiple independent networks and (2) balance minimizing messaging overhead and avoiding wasted contact times in bandwidth-constrained sensor internetworks. Mechanisms to accomplish these goals in the context of a CSI are absent in the literature.

3 Related Work

We survey three types of related work: quality of service (QoS) in DTNs, message aggregation in store-and-forward networks, and the use of contact graphs for overlay routing.

3.1 QoS in Heterogeneous Challenged Networks

The term Quality of Service (QoS) refers to a variety of objective and subjective measurements relating to the usability of data received through a network. In the context of any network architecture, a Service-Level Specification (SLS) may objec-tively define QoS [4]. An SLS, itself, is a set of parameters whose values define the service offered to a traffic stream by a domain [5]. These parameters vary based on the type of network and the protocols and applications using the network. The parameters used to objectively measure QoS in low-latency networks will be different from those used to measure QoS in challenged/disrupted networks.

Research in the area of QoS for heterogeneous networks focuses heavily on selecting between existing QoS services in a given local network (horizontal handoff) or pushing traffic to different networks with other QoS services (vertical handoff) [4].

While the literature is dense with algorithms for horizontal handoffs in support of QoS for low-latency networks, vertical handoffs, which are required in heterogeneous networks, remain an active research area. Approaches to vertical handoffs involve omniscient resource managers and constant end-to-end path monitoring [6] and in general these schemes require sophisticated handover management algorithms, cost functions, and performance evaluation [7]. None of these schemes are appropriate for CSIs with intermittent connectivity as link disruptions prevent the useful collection of information to make these handoff decisions.

In a challenged network end-to-end delivery is not guaranteed, so any QoS mechanism can only be applied at the interface between the application and the network. In addition to best-effort and end-to-end provisioning service classes, a DTN-specific set of service classes focused on allotted bandwidth has been proposed [8]. Allotted bandwidth captures a requested data rate over an interval corrected for link delay, periodic availability, and other latencies. The benefit of this approach is that it informs the queuing limits necessary to communicate a message through a store-and-forward network [8]. Within a given service class there are proposals for priority classes (low, medium, high) and message delivery notifications that could inform scheduling, flow, and congestion control options [9] for the local network.

Consistent with allotted bandwidth service classes, DTN deployments for space networks focus on traffic shaping in accordance with one of four policies: isochronous for real-time data, minimum rate for applications that can tolerate jitter, bulk with deadline for accumulating file transfers, and best-effort [10]. Research on internetworking space networks remains largely focused on dedicated relay nodes and gateways [11].

We discovered no algorithms for the application of these techniques to graph-routed internetworks operating absent service gateways and abstracted from the underlying physical properties of constituent networks.

3.2   Message Aggregation and Fragmentation

Message fragmentation occurs either as a reaction to link congestion or as part of a pro-active flow control mechanism [12]. Reactive fragmentation is typically preferred to pro-active fragmentation in low-latency networks because it results from the instantaneous characteristics of the transmission link, rather than relying on a hard-coded configuration value [13]. In technical demonstrations, both proactive and reactive fragmentation strategies have successfully passed large data across a series of links such that the entire data set could not be communicated over a single link [14]. As the latency of the network increases, the ability to meaningfully measure the state of link congestion decreases making proactive fragmentation a more plausible option. The ability to pro-actively fragment a message based on some network information improves message delivery ratios, as long as unnecessary fragmentation in the network can be avoided [15].

DTN protocols, such as the Bundle Protocol [16], provide mechanisms for fragmentation. These mechanisms typically react to a next-hop bottleneck and are not, alone, appropriate for tuning system-level parameters such as maximizing use of small residual contact capacity. Additionally, these protocols do not implement in-network data fusion, instead relying on users themselves to combine their payload data.

While there is significant literature devoted to reactive fragmentation algorithms, there is very little work on improving proactive fragmentation techniques. Specifically, we have found no work that performs path validation with the purpose of inferring optimal per-path routing fragments. Further, we have not encountered any research performing the inverse function: proactive aggregation at the networking layer when message sizes are small compared to available bandwidth.

3.3   Contact-Graphs for Overlay Routing

Pre-configuring contact opportunities into a CG is a community-supported strategy for planning communications in space networks [17–19] where topological change is deterministic. Contacts in the CG represent the physical bandwidth capacity between two network nodes over time and graph-theoretic algorithms, such as Dijkstra's algorithm [20] and the Contact Graph Routing Protocol (CGR) [21], may be used to build a message path as part of a routing function. CSIs can either preconfigure contacts or infer them from data-link and transport layers of individual links. CSI nodes may synchronize this information over common messaging paths across the internetwork [3]. A concise definition of the CG and its contacts is as follows.

Definition 1. A Contact Graph $CG = (V, E)$ is a directed, weighted graph of contact opportunities, E, amongst nodes, V, in a network, over time. Each node, i, in the network has its own local graph, CG_i. A contact between nodes a,b in a graph is represented by the capacity function $a^b CG_i \in \Re$. Positive values indicate transmission capacity. In the context of a node, we refer to a contact as $C_{a,b} = a^b CG_i$.

Definition 2. The initial transmission capacity (bandwidth) of a contact is measured as the number of bytes that can be received over the contact. Bandwidth (C_{BW}) is initialized as a function of the start and end times (C_{START}, C_{END}), data rate (C_{RATE}), and one-way propagation delay (C_{PROP_DELAY}) of the contact, as follows:

$$C_{BW} = C_{RATE*}(C_{END} - C_{START} + 1 - C_{PROP\_DELAY})$$

Some implementations of contact-based routing use the local CG to calculate only the next-hop of a plausible message path [21]. However, an entire message path may be calculated and added to a message as a routing header. This routing header may be evaluated at each waypoint and updated when necessary. In cases where CGs differ amongst nodes in the network, routing headers both prevent loops and enable a wider variety of routing cost functions [22].

4 Quality of Service in CSIs

CSIs cannot provide end-to-end rate-based guarantees as end-to-end connectivity may not exist at any given point in time and active links through the internetwork may have different bandwidths, data rates, and utilization. The concepts of allotted bandwidth, traffic shaping, and bulk delivery in accordance with specified deadlines do remain relevant to CSIs where sensed data need only be available for asynchronous consumption and not part of a real-time or other synchronous data stream.

We make two observations regarding the nature of message delivery within a CSI. First, data sources and sinks are eventually connected by one or more contacts over time. Second, message delivery within a graph-based routing approach is deadline-based: data transmission is only guaranteed to occur such that data can be received prior to the end of the contact. From these observations, we define new terms to describe DTQoS in these networks.

Definition 3. The rate measurement "bits per active second" (*bpas*) identifies the number of bits transmitted from a source to a sink over a DTN/CSI only considering times when links along the path exist. The difference between bits per second (*bps*) and *bpas* is illustrated in Fig. 3. In a CSI, the *bpas* is the amortized bits per second over the duration of the link.

Definition 4. The rate measurement "bits per link" (*bpl*) identifies the number of transmitted bits receivable over a link, assuming no impairments. *Bpl* is less than or equal to the total bandwidth available over a link. It is defined as the *bpas* * the duration of the link, in seconds. For example, a 10 s contact supporting a 30 kbps *bpas* would need to reserve 300 kb of link bandwidth.

Within the context of the CSI architecture, we use a bandwidth allotment mechanism expressed as a *bpas* and enforced on each link as a minimum *bpl*. In other words, so long as there is physical connectivity in the network, DTQoS will provide message delivery in accordance with these new rate measurements. Enforcement becomes a matter of traffic shaping where each node is allowed access to only a portion of the

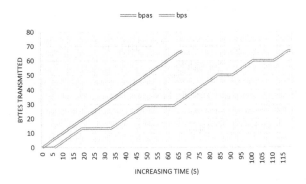

Fig. 3. A performance rate can be defined based on active seconds (left) rather than absolute seconds (right).

physical link. In such a scenario, we can re-draw Fig. 2 as Fig. 4 below. Notably, the *order* in which these messages are transmitted during the contact is irrelevant. The only guarantee is that all of the messages that are queued for the next contact are transmitted prior to the end of the contact.

In this example, *bpl* allocations divide the next link based on configured *bpas* allocation. Specifically to the example in Fig. 4, each of the three tier-1 networks transmit at least one message into the next link whereas in Fig. 2 network S3 would have had no transmission opportunity.

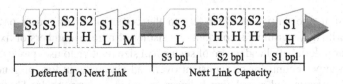

Fig. 4. A prioritized transmission queue ensures that applications using a contact in an internetwork are not starved for bandwidth.

4.1   The Traffic-Shaping Contacts Method

Contacts in a CG are the logical representations of physical links. By slightly altering the definition of a contact, we can represent the *bpl* of the link with respect to a particular user class and SLS rather than the entire link bandwidth. We note that the specification of a particular *bpl* and SLS is negotiated amongst network and internetwork stakeholders as part of network configuration and not automatically calculated on the fly by the network. Algorithms for dynamically altering *bpl* allocations based on local network state is an area for future work.

We augment contact information with a *Unique Identifier* (UID) for a particular SLS such that for the entire bandwidth capacity over a temporal physical link, multiple contacts are defined, one per UID. A reserved UID (*0*) indicates generally available bandwidth, such as that used for retransmission margin or as a reservation pool from which to allocate bandwidth for new UIDs in the future. The relationship between link capacity (L_{cap}) and contact capacity is captured in Eq. 1, where a and b represent the nodes of a contact, u represents the UID identifying a particular service-level specification, and $C_{a,b,u}$ represents the traffic-shaping contact capturing the allotted bandwidth for user class u over contact $C_{a,b}$.

$$L_{cap} = \sum_{u=0}^{u=n} C_{a,b,u} \qquad (1)$$

When a message associated with a particular service-level specification is ready for routing at Node i, the set of contacts tagged with the UID for that specification comprises an independent CG, $CG_{i,u}$, defined in Eq. 2.

$$CG_{i,u} = \langle C_{a,b,u} \cup C_{a,b,u0} | a, b \in V \rangle \qquad (2)$$

In this instance, when a message, m, for the service-level specification identified by u is to be routed in the system, the graph-based routing algorithm at node i uses $CG_{i,u}$ rather than CG_i. In doing so, traffic shaping occurs as a natural consequence of honoring contact capacity. This mechanism does not specify *when* in the course of a contact a particular message will be transmitted, but does guarantee before the end of the contact that there will be a transmission opportunity.

5   Message Fragmentation and Aggregation

In this section, the term message refers to the combination of a user payload and overhead, which we assume is constant, consisting of regular numbers of fixed-length headers. Variable-length headers are of such small size relative to user payloads that we estimate an upper-bound for our constant overhead size without loss of generality.

There is no required correlation between message volume, payload size, and *bpl* allocations. Applications may fill their *bpl* allocations with many fixed-sized messages, with bursts of large messages, or with random message sizes driven by the underlying sensed phenomenology. In all cases, it is unlikely that a set of messages will exactly consume a configured *bpl* allocation; referring back to Fig. 4, we see sections of unused bandwidth within each *bpl* allocation.

For any feasible path through the network, at a given point in time, we can calculate a desirable payload size that balances minimizing wasted bandwidth and minimizing messaging overhead. Nodes in the network calculate this size using information from their local CG and then use this size as the basis for aggregation and fragmentation algorithms. Since contacts in the CG remember their residual capacity, a *Path Preferred Size* (PPS) may be calculated as the smallest residual capacity of any contact comprising the path, less a constant overhead margin, to account for messaging overhead from the physical links and routing headers.

While we desire to keep fragments as large as possible, there is a practical concern relating to the largest fragment allowed in the overlay. Message fragments may be limited by Maximum Transmission Units (MTUs), time-division multiplexing within a contact duration, and duplex operation of a transceiver, all of which may prevent an entire contact opportunity being used for transmission. In such cases, we support a configured maximum payload size which can be lower than the calculated PPS value. Going forward, the term PPS refers to the smaller of this configured size, or the calculated PPS value.

The problem of fitting a series of variable-sized payloads into a calculated PPS is similar to the multidimensional 0/1 Knapsack problem [23], where individual message payloads are the series of knapsack items, $x_1, x_2, \ldots, x_n$, and the knapsack capacity is the overall PPS value. This knapsack problem is known NP-Complete and polynomial approximations remain complex in time and memory. We recommend for embedded systems the use of greedy approximation solutions that first sort messages by value function and then add them to the transmission queue from largest to smallest value [24]. By defining value as the ratio of payload data over *bpl* it is clear that the highest value comes from the largest payload, as it will incur the least amount of messaging overhead. Therefore, finding (or constructing) a payload that exactly matches the PPS represents the optimal use of the current path.

In the remainder of this section we present first an aggregation approach, a message fragmentation approach, and the PAF algorithm that combines them.

5.1 Message Aggregation

If the *PPS* is larger than the current payload size (P_{BYTES}), the sending node has the opportunity to merge several payloads into one aggregate payload (AP). Payloads are candidates for aggregation if they share a destination, UID, associated security mechanism, and any other common attributes that allow a path for one payload to work for all aggregated payloads. The bit savings of a specific grouping of like-payloads, *S*, of the aggregation approach is measured by the *future* bandwidth freed by the aggregation. This is the sum of the contact bandwidth used (and thus not needed for transmitting over future contacts) and the reduction in message overhead, calculated in Eq. 3.

$$S = (AP_{BYTES} - P_{BYTES}) + O * (AP_{MSGS-1}) \tag{3}$$

Where O is the per-message overhead, AP_{BYTES} is the size of the aggregated payload, and AP_{MSGS} is the number of whole payloads aggregated together and thus not requiring per-payload messaging overhead.

When AP_{BYTES} equals *PPS* we have an optimal result: we have maximized transmission opportunities by using all practical bandwidth along the path and we have minimized message overhead to the overhead of one message. In most cases, however, $AP_{BYTES} < PPS$ and we must decide whether to ignore the delta bandwidth or to fragment a payload to fill this remainder.

5.2 Message Fragmentation

Once the aggregation algorithm has finished, there may still exist some estimated residual capacity (*RC*) left in the AP such that $0 < RC < PPS$. A decision must be made as to whether another payload should be fragmented and used to completely fill the *AP* prior to its transmission.

There are two reasons to fragment a payload at the overlay. First, we avoid wasting capacity as, at the time we perform the calculation, there is no known payload that can fit within *RC*. If no such payload is generated by the application prior to the contact expiration then *RC* bytes of transmission opportunity are wasted. Second, in scenarios where a payload exceeds a given *bpl* allocation, then fragmentation must be performed else the payload will never be transmitted. Conversely, there is a reason to not fragment a payload at the overlay: if the additional overhead burden incurred by fragmentation is greater than the *RC* being saved. This burden increases with the number of fragments generated, because each fragment will incur its own messaging overhead.

Definition 5. The burden, *B*, associated with fragmenting a payload, *P*, is the difference between the total bytes generated from payload fragmentation based on

RC ($Size_{Cur}$) and the total bytes generated from payload fragmentation based on the overall *bpl* allocation ($Size_{Best}$). When calculating $Size_{Cur}$, this first payload fragment (P_{RC}) is made large enough to fill RC (accounting for message overhead) leaving the remaining payload, P-P_{RC}, to be communicated during a future contact.

$$Size_{Best} = \left\lceil \frac{P}{(bpl - O)} \right\rceil * O + P \tag{4}$$

$$P_{RC} = P - (RC - O) \tag{5}$$

$$Size_{Cur} = RC + \left\lceil \frac{P_{RC}}{(bpl - O)} \right\rceil * O + P_{RC} \tag{6}$$

$$B = Size_{Cur} - Size_{Best} \tag{7}$$

When $B >= RC$, we are using as much or more bandwidth than we are saving by fragmentation. For every payload for which $B < RC$, we select the payload whose fragmentation minimizes B. We then add $(RC-O)$ bytes of the payload to AP, and the remaining P_{RC} back into the payload message queue.

When calculating the burden associated with fragmentation we do not directly factor in the link layer overhead. This value is captured as part of the configurable overhead adjustment, O.

5.3   The PAF Algorithm

The Payload Aggregation and Fragmentation process is illustrated in Fig. 5. In this figure, the three major processing chains are grouped within boxes and labeled 1, 2, and 3. These processing chains operate as follows:

1. The typical process of passing a message into a routing module to receive the message path is replaced by a probe message. This probe message contains the minimum size of a message (denoted by the message overhead value O) and the SLS identifier of the message (the *UID* value). From this path, a PPS value is calculated and the RC of the path is set to this value. Based on the size of the payload versus the RC value the algorithm will aggregate or fragment.
2. The payload is aggregated by including it with the AP. Once this has been done, another payload is retrieved from the payload queue, the RC value is updated to reflect the new AP size, and the algorithm decides whether to further aggregate or fragment.
3. The payload fragmentation burden, B, is calculated and used to decide whether the payload should be fragmented or whether it is better to simply waste the contacts RC capacity. If $B < RC$ then fragmentation should be used and the payload fragment is added to the AP and the remainder is put back to the payload queue. In either case, the fragmentation step ends the algorithm and the AP is passed to the transmission queue for that contact.

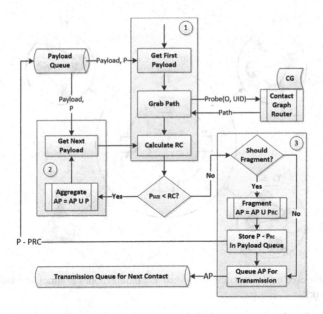

Fig. 5. PAF assembles an aggregate payload from a queue of payloads.

Elided from this algorithm is the more trivial case that if the payload queue is exhausted of payloads, then the AP to date is simply queued for transmission.

6 Simulation and Results

To demonstrate the performance of TSC and PAF in a graph-routed network, we build a representative topology and associated simulation using the ns-3 network simulator. Figure 6 illustrates a federated sensor internetwork in accordance with the architecture presented in Sect. 2 and Fig. 2. Three tier-1 networks (T1.1, T1.2, T1.3) produce data and prioritize them according to local schemes. A tier-2 data collection network consisting of three regionally mobile nodes (T2) provides regular passes through the tier-1 networks to capture data and ferry them to a tier-3 exfiltration node (T3).

The data volume and periodicity of the tier-1 networks are designed to simulate a variety of low-rate monitoring data from sensor cluster heads. T1.1 produces streaming video at 32 kbps, a lower bound for compressed H.264 video. T1.2 produces a bulk file transfer of 1 MB every 7 s. T1.3 produces mixed payloads: a constant 8 Kbps low-priority audio feed, a 45 s half-frame-rate medium priority 16 kbps video feed every minute, and a high priority 4 MB photo every 5 min. These data volumes and production rates simulate the type of competing, variable priority traffic experienced in a federated, resource constrained sensor internetwork. There is no relative priority between T1.1, T1.2, and T1.3; from the point of view of the internetwork, each local network must communicate its data with equal priority.

The T2 mobile nodes follow each other in a circular pattern and take 30 min to complete a circuit. Based on geometry, each sensor network sees a T2 mobile node for

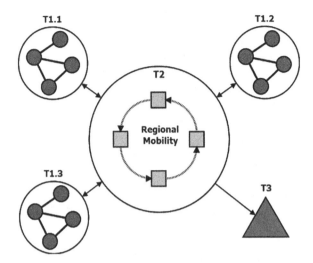

Fig. 6. A multi-tiered ns-3 simulation uses fixed and mobile nodes to simulate a variety of traffic in an internetwork.

approximately 15 out of every 30 min, with each link lasting approximately 5 min. Links in the system are high rate up to 10 Mbps, and we impose a 60 KB limit on the size of any user message payload.

To demonstrate the value of traffic-shaping contacts and message aggregation, we run the simulation in two configurations: one with sufficient *bps/bpl* allotments to account for all user traffic, and again with insufficient allotments. These allotments are given in Table 1. Running the simulation with nominal rates simply results in all data being delivered which is as expected in a network with no challenges for resources. As a constrained network, the regular, small packets produced by T1.1 and T1.3 streaming protocols will need to be managed to preserve bandwidth allocation for the larger, more periodic file transfers from T1.2.

Only the *bpl* allocations are allowed for internetwork data transmission with the remaining bandwidth assumed reserved for other, local network traffic. We run the network with and without TSC/PAF. When running with TSC/PAF, three versions of each contact are defined, labeled with the appropriate T1.1/T1.2/T1.3 UID, and assigned the *"Configured bpl"* from Table 1. Running without TSC/PAF there is a single contact between nodes and no associated UID, and the capacity of the contact is the sum of the three *bpl* values.

Table 1. DTQoS analysis and configuration captures the SLS associated with sensor networks in the TSC/PAF simulation.

Sensor network	bps (Kbps)	Nominal bpas (Kbps)	Nominal bpl (KB)	Configured bpl (KB)
T1.1	32	64	2,400	183
T1.2	1,200	2,400	90,000	5,188
T1.3	136	272	35,116	9,394

For each topological configuration we examine message delivery rate, the ratio of payload bytes sourced verses delivered, and overall bytes delivered.

6.1   Message Delivery Rate

We examine the receipt time of messages at the tier-3 exfiltration node as an indication of how the tier-2 mobile nodes have been populated. As previously stated, the data prioritizations within individual tier-1 networks are not assumed to be comparable at the tier-2 data collection nodes. We expect that the application of the TSC method alters the type and order of message delivery over a delivery with no traffic shaping. Further, we expect that the PAF algorithm will result in fewer messages as payloads are aggregated.

Figure 7 illustrates the message delivery to the tier-3 exfiltration node over time, by originating tier-1 network. Early in the simulation, data are not communicated due to the geometries of the internetwork and the data production. In this figure, message delivery resembles a step function where messages are bulk delivered at the start of contact between the T2 and T3 nodes, as the T2 nodes utilize the full bandwidth of the link without any traffic shaping. Additionally, the T1.1 network, which produces the smallest data volume in the simulation produces and delivers almost half of all messages.

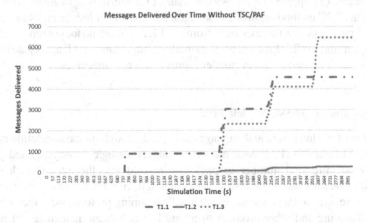

Fig. 7. Without TSC/PAF the T1.1 network, which produces the smallest data volume, dominates message delivery as it sees the data mules first.

Figure 8 illustrates the same data run as Fig. 7, but using the TSC/PAF algorithms. There is a 77 % reduction in the number of T1.1 messages, and a 23 % size reduction in the number of T1.3 messages achieved with using TSC/PAF. The large size reduction for T1.1 messages is based on the ability to aggregate the several small messages that make up its 32 kbps stream. The smaller reduction for the T1.3 messages is due to the periodic 4 Mb file transfer from T1.3 which generates very large payloads that cannot be aggregated. Notably, there is a 200 % size *increase* in the number of messages from T1.2. Without TSC/PAF, the large and periodic T1.2 file messages are overwhelmed by the higher-rate producing T1.1 network. Using TSC/PAF the T1.2 *bpl* allocations are preserved and we are able to receive more of those messages.

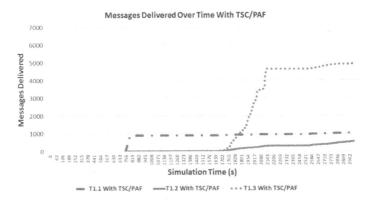

Fig. 8. With TSC/PAF message delivery is even across the three tier-1 networks and accommodates burst bulk file transfers from T1.2.

Figure 8 shows the following three benefits of TSC/PAF. (1) Individual data deliveries are throttled over time as illustrated by smoothed delivery rates (versus the "step function" appearance of Fig. 7) which shows that the internetwork is honoring configured *bpl* values. (2) Approximately twice as many T1.2 data messages are delivered in the system as the TSC method ensures a more equitable sharing of link capacity amongst the three tier-1 networks. (3) The messages from the T1.1 network no longer represent half of all message traffic. As T1.1 represents the smallest data volume, PAF aggregates the high number of small messages into a smaller number of larger messages.

6.2   Payloads vs. Messages Delivered

The number of payloads sent and messages delivered throughout the simulation is shown in Fig. 9. Without TSC/PAF each message contains a single non-aggregated payload and, therefore, the total number of payloads injected into the internetwork (11,299 payloads) is the same as the number of messages delivered by the internetwork (11,299 messages). Several of these messages are small streaming protocol messages where the associated routing and other protocol overhead is a significant percentage of message size. Combining payloads allows PAF to communicate more payload information using fewer messages in the network. With PAF 23,053 payloads are aggregated into just 6,543 messages. This increase in payloads is enabled by the lower overhead associated with having fewer messages and, thus, more capacity to communicate user data.

6.3   Bytes Delivered

Figure 10 illustrates the additional goodput through the internetwork associated with the TSC/PAF algorithms. Without TSC/PAF, a total of 42.39 MB of user data are received through the internetwork, versus 68.92 MB. Our algorithms increased goodput by approximately 63 %. From Fig. 9 we recall that this goodput increase comes with an overall 43 % reduction in the number of messages in the system.

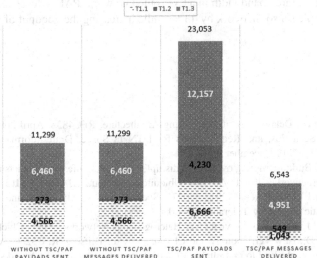

Fig. 9. When using TSC/PAF applications are able to send more payloads in fewer messages.

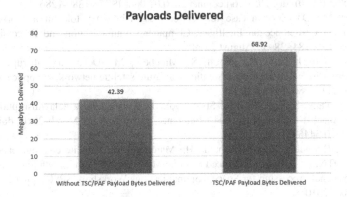

Fig. 10. TSC/PAF efficiencies enable more data to be communicated across the internetwork.

7 Conclusion

Challenged sensor internetworks (CSIs) are supersets of DTNs that support sufficient determinism to build and maintain a CG along commonly used message routes. These internetworks require a minimum quality-of-service provisioning and enforcement mechanism to prevent application data from being passed over for transmission and to limit the amount of internetwork traffic injected into any constituent local network. We present TSC, a method for allowing multiple contacts to share a physical link for the purposes of traffic shaping. In conjunction with TSC, we propose new terms for discussing QoS agreements in CSIs and DTNs, as rate guarantees during periods of link connectivity. We further present PAF, an algorithm for aggregating multiple messages based on preconfigured bandwidth in a CG-routed network. PAF reduces the amount of traffic in a sample sensor network by 43 % while increasing the goodput of the network by 63 %.

References

1. Cerf, V., et al.: Delay-Tolerant Networking Architecture, RFC4838, April 2007
2. Rationale, Scenarios, and Requirements for DTN in Space, Draft Informational Report, CCSDS 734.0-G-0, December 2009
3. Birrane, E.: Building routing overlays in disrupted networks: inferring contacts in challenged sensor internetworks. Int. J. Ad Hoc Ubiquitous Comput. (IJAHUC) **11**(2–3), 139–156 (2012). doi:10.1504/IJAHUC.2012.050271. Special issue on Algorithms and Protocols for Opportunistic and Delay Tolerant Networks
4. Marchese, M.: Quality of Service Over Heterogeneous Networks. Wiley, Chichester (2007)
5. Grossman, D.: New Terminology and Clarifications for Diffserv, RFC3260, April 2002
6. Dugeon, O., et.al.: End to end quality of service over heterogeneous networks: EuQoS. In: Proceedings of NetCon 2005, Lanion, France, November 2005
7. Yan, X., Şekercioğlu, Y.A., Narayanan, S.: A survey of vertical handover decision algorithms in fourth generation heterogeneous wireless networks. Comput. Netw. **54**(11–2), 1848–1863 (2010). doi:10.1016/j.comnet.2010.02.006. ISSN: 1389-1286
8. Demmer, M.: DTNServ: a case for service classes in delay tolerant networks. In: 4th International Conference on Intelligent Computer Communication and Processing, ICCP 2008, pp. 177–184, 28–30 August 2008
9. Caini, C., Cruickshank, H., Farrell, S., Marchese, M.: Delay- and disruption-tolerant networking (DTN): an alternative solution for future satellite networking applications. Proc. IEEE **99**(11), 1980–1997 (2011)
10. Tsao, P., Wang, S.-Y., Gao, J.L.: Space QoS framework over a delay/disruption tolerant network. In: 2010 IEEE Aerospace Conference, pp. 1–5, 6–13 March 2010. doi:10.1109/AERO.2010.5446951
11. Caini, C., Firrincieli, R., Cruickshank, H., Marchese, M.: Satellite communications: from PEPs to DTN. In: 2010 5th Advanced Satellite Multimedia Systems Conference (Asma) and the 11th Signal Processing for Space Communications Workshop (SPSC), pp. 62–67, 13–15 September 2010
12. Fall, K., Farrell, S.: DTN: an architectural retrospective. IEEE J. Sel. Areas Commun. **26**(5), 828–836 (2008)
13. Magaia, N., Pereira, P.R., Casaca, A., Rodrigues, J.J.P.C., Dias, J.A., Isento, J.N., Cervello-Pastor, C., Gallego, J.: Bundles fragmentation in vehicular delay-tolerant networks. In: 2011 7th EURO-NGI Conference on Next Generation Internet (NGI), pp. 1–6, 27–29 June 2011. doi:10.1109/NGI.2011.5985945
14. Ivancic, W.D., Paulsen, P., Stewart, D., Eddy, W., McKim, J., Taylor, J., Lynch, S., Heberle, J., Northam, J., Jackson, C., Wood, L.: Large file transfers from space using multiple ground terminals and delay-tolerant networking. In: 2010 IEEE Global Telecommunications Conference (GLOBECOM 2010), pp. 1–6, 6–10 December 2010. doi:10.1109/GLOCOM. 2010.5683304

15. Pitkanen, M., Keranen, A., Ott, J.: Message fragmentation in opportunistic DTNs. In: 2008 International Symposium on a World of Wireless, Mobile and Multimedia Networks, WoWMoM 2008, pp. 1–7, 23–26 June 2008. doi:10.1109/WOWMOM.2008.4594892

16. Burleigh, S., Scott, K.: Bundle Protocol Specification, November 2007. http://tools.ietf.org/html/rfc5050

17. Sekhar, A., et al.: MARVIN: Movement-Aware Routing oVer Interplanetary Networks. In: IEEE SECON (2004)

18. Wyatt, J., et al.: Disruption tolerant networking flight validation experiment on nasa's epoxi mission. In: 2009 First International Conference on Advances in Satellite and Space Communications, pp. 187–196 (2009)

19. Caini, C., Firrincieli, R.: Application of contact graph routing to LEO satellite DTN communications. In: 2012 IEEE International Conference on Communications (ICC), pp. 3301–3305, 10–15 June 2012

20. Segui, J., Jennings, E., Burleigh, S.: Enhancing contact graph routing for delay tolerant space networking. In: IEEE GLOBECOM (2011)

21. Burleigh, S.: Contact Graph Routing: draft- burleigh-dtnrg-cgr-01, July 2010. http://tools.ietf.org/html/draft-burleigh-dtnrg-cgr-01

22. Birrane, E.: Improving graph-based overlay routing in delay tolerant networks. In: Proceedings of IFIP Wireless Days (2011)

23. Fréville, A.: The multidimensional 0–1 knapsack problem: an overview. Eur. J. Oper. Res. 155(1), 1–21 (2004). doi:10.1016/S0377-2217(03)00274-1. ISSN: 0377-2217

24. Dantzig, G.B.: Discrete-variable extremum problems. Oper. Res. 5(2), 266–288 (1957)

Virtualbricks for DTN Satellite Communications Research and Education

Pietrofrancesco Apollonio[1], Carlo Caini[1(✉)], Marco Giusti[1],
and Daniele Lacamera[2]

[1] DEI-ARCES, University of Bologna, Bologna, Italy
f.apollonio@ldlabs.org, carlo.caini@unibo.it,
marco.giusti3@studio.unibo.it
[2] TASS Technology Solutions, Leuven, Belgium
daniele.lacamera@tass.be

Abstract. Virtualbricks is a virtualization solution for GNU/Linux platforms developed by the authors and included in Debian. The paper aims to show its potential, referring to version 1.0, just released, when applied to both research and education on DTN satellite communications. In brief, Virtualbricks is a frontend for the management of Qemu/KVM Virtual Machines (VMs) and VDE virtualized network devices (switches, channel emulators, etc.). It can be used to manage either isolated VMs, or testbeds consisting of many VMs interconnected by VDE elements. Among the wide variety of possible applications, with or without VM interconnections, the focus here is on the development of a virtual testbed on DTN satellite communications, a task for which Virtualbricks was especially designed. After having introduced the main characteristics of Virtualbricks, in the paper we will show how to set-up a Virtualbricks testbed, taking as an example a testbed recently used by the authors to investigate Moon communications through orbiters. The validity of Virtualbricks results is confirmed by comparison with results achieved on a real testbed, set-up for this purpose. The same testbed has also been successfully used for educational purposes at the University of Bologna.

Keywords: Testbed virtualization · DTN · Satellite communications · KVM · Qemu · VDE

1 Introduction

Several complete solutions for virtualization management have been developed to simplify design, configuration and management of Virtual Machines (VMs). Currently available virtualization management tools include proprietary suites like VMware® [1], Microsoft® SCVMM [2], Paragon® VM [3], SolarWinds® [4] and many more. Other solutions are based on Free or Open Source software and in particular on the use of the Kernel based Virtual Machine (KVM) [5], which is now part of the Linux kernel. Unfortunately, none of these solutions focuses on the design and management of complex network layouts, as they were created for cloud management and the virtualization of enterprise server farms does not usually require complex topologies.

© ICST Institute for Computer Sciences, Social Informatics and Telecommunications Engineering 2016
I. Bisio (Ed.): PSATS 2014, LNICST 148, pp. 76–88, 2016.
DOI: 10.1007/978-3-319-47081-8_7

There are few examples in the literature that exploit virtualization technologies to provide virtual equivalent of testbed components, like routers and switches. The most advanced is possibly provided by Marionnet [6, 7]. Marionnet uses Virtual Distributed Ethernet (VDE) [8, 9], to emulate the network infrastructure needed to interconnect the VMs in complex networks. Unfortunately, Marionnet is mostly intended for educational purposes and the design choice to implement VMs via User-mode Linux (UML) [10] greatly penalizes performance. A different approach to build network testbeds is followed by Mininet [11], which allows the user to run a collection of end-hosts, switches, routers, and links on a single Linux kernel. It uses lightweight virtualization to make a single system look like a complete network, running the same kernel, system, and user code. It is however a powerful network emulator, not a VM manager. In particular, its nodes do not run independent OSs and must share the file system of the host.

Since we found no existing solutions that would meet our requirements, we started the development of Virtualbricks [12], with the aim of creating a virtualization management tool focused on network design. Virtualbricks relies on Qemu [13], KVM and VDE; it is released under the GNU General Public License version 2 and has been included in the GNU/Linux Debian distribution. The paper refers to the version 1.0, just released, which implied an in-depth redesign of many part of the code to improve stability and add new significant features. Virtualbricks stems from the fusion of Qemulator, a front-end for Qemu, and Virtual NetManager, a project previously developed by the authors. The aim of the project is to provide an easy interface not only for VM management, but also for virtual network design. Although not exclusive, one of the most significant applications is the virtualization of real testbeds, such as those used by the authors [14] to evaluate TCP and DTN (Delay-/Disruption- Tolerant Networking) protocols in challenged networks [15]. As well as for research, Virtualbricks has a great potential also as an educational tool, as it allows students to build a complete networking testbed on their own PCs. Many new features introduced in the release 1.0 derive from the practical experience gained by the use of release1.0 beta versions in the LAB activities of a TLC Engineering Master course of the University of Bologna [16].

This paper aims to present the main features of Virtualbricks seen from the user point of view. Although many applications do not require VM interconnections, here the focus is on the development of a virtual testbed for scientific research on satellite DTN communications, a task for which Virtualbricks was especially designed. To this end, after a general description of Virtualbricks, the steps necessary to build the virtual testbed that was used by the authors in [17] are presented. The accuracy of Virtualbricks results will be then validated by comparisons with those achievable with a real testbed, expressly built up for this purpose. It is however worth stressing that Virtualbricks is a front-end, thus performance achievable (in terms of accuracy and speed) is the same as that achievable by its components, KVM or Qemu for VMs, and VDE for virtual network devices. The advantages of Virtualbricks stem from the integration of these components in a flexible and powerful interface. For example, a complete testbed consisting of many interconnected VMs, with their own file systems, can be saved in just one tar file and exported to other PCs or made available for download on a web site.

2 Virtualbricks General Description: "Main" Window and "Bricks"

In this section we will first examine the Virtualbricks main window, to give a general idea of Virtualbricks features; then we will focus on the "bricks", i.e. the VMs and the VDE elements that are at the core of each project, and on their configuration.

2.1 "Main" Window

The Main window is the operating centre of Virtualbricks. There are five different pages and several command menus (Fig. 1). "Bricks" is the first page and, once selected, it shows an icon for each brick of the current project. A new brick can be added by pressing the "New brick" button and then by selecting the wanted brick from the selection page (Fig. 2).

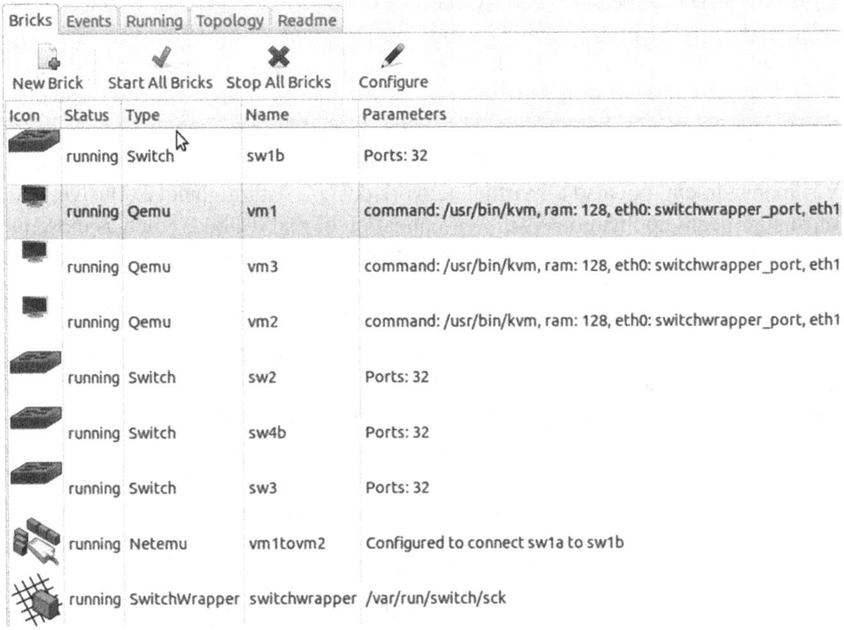

Icon	Status	Type	Name	Parameters
	running	Switch	sw1b	Ports: 32
	running	Qemu	vm1	command: /usr/bin/kvm, ram: 128, eth0: switchwrapper_port, eth1
	running	Qemu	vm3	command: /usr/bin/kvm, ram: 128, eth0: switchwrapper_port, eth1
	running	Qemu	vm2	command: /usr/bin/kvm, ram: 128, eth0: switchwrapper_port, eth1
	running	Switch	sw2	Ports: 32
	running	Switch	sw4b	Ports: 32
	running	Switch	sw3	Ports: 32
	running	Netemu	vm1tovm2	Configured to connect sw1a to sw1b
	running	SwitchWrapper	switchwrapper	/var/run/switch/sck

Bricks | Events | Running | Topology | Readme
New Brick | Start All Bricks | Stop All Bricks | Configure

Fig. 1. The Home window of Virtualbricks.

Once selected, an existing brick can be powered on/off, configured, deleted and renamed. The second page of the main window is "Events". It is used to set commands (e.g. a script) that can be executed by Virtualbricks in an autonomous way. The third page, "Running", lists the running bricks. As in the first page, a number of operations can be performed on the brick selected (opening and closing the configuration terminal, sending ACPI signals, killing processes, etc.). The "Topology" page shows an

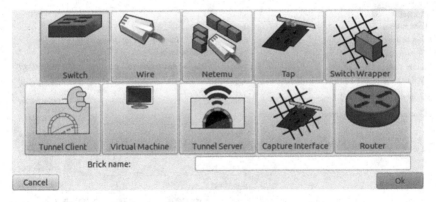

Fig. 2. The Brick selection page of Virtualbricks.

interactive image of the entire topology, which can be useful for visually checking the brick connections. The topology layout can also be exported as an image file. The latest page "Readme", allows either reading or writing comments on the testbed. It can be useful when a testbed is distributed to other researchers or to students.

Virtualbricks supports two kinds of virtual machines: Qemu and KVM. Qemu [13] is a processor emulator and supports a large variety of guest OS, including those built for CPU architectures that differs from that of the host machine. KVM [5] is a virtualization solution for x86 processors based on hardware virtualization technologies (Intel Virtualization Technologies® and AMD-Virtualization®). The full virtualization approach of KVM, when compared to CPU emulators, offers many advantages in terms of guest performance and access to paravirtualized devices, such as virtual disks and virtual Ethernet tools. Moreover, scalability is greatly improved by the joint use of KVM and Kernel Samepage Merging (KSM) [18], a Linux feature that combines multiple identical memory pages that are in use by different processes into a single physical RAM page ("overcommit" feature). On the other hand, KVM needs a CPU with the virtualization extensions and does not support binary translation [19]. Due to their superior performance, KVM is generally preferable whenever there is no need to emulate a different architecture.

2.2 VM Brick Configuration

The VM brick configuration panel can be opened by clicking on the brick with the right mouse button and selecting "Configure". The VM configuration options are grouped into different pages, each representing a different section. The "Drives" page (Fig. 3) enables the configuration of all the block devices, in the form of either virtual disk images or direct access to host peripherals such as CD-ROMs. Disk images cannot be used concurrently with write-on permission by more than one VM, to avoid the corruption of image content. To circumvent this problem, Virtualbricks stores local images in a database. Virtual disk images can be managed directly from Virtualbricks. In particular, a differential disk image can be used by checking the "Private Cow" box.

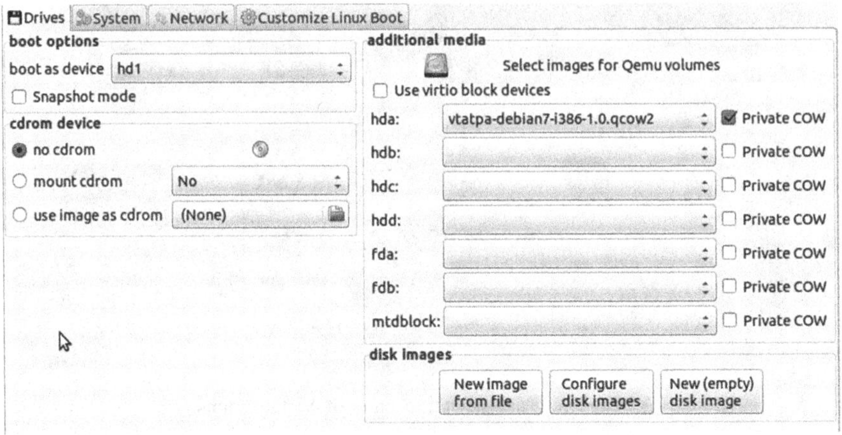

Fig. 3. The Drives configuration page of a VM brick.

This enables the use of only one single image for all VMs, thus greatly improving the consistency and the dimension of the project. Cow files are used in this case only to save machine dependent configurations. Another important check box is "Snapshot Mode". If checked, no changes are saved when the VMs are switched off, which can be useful in an educational Lab or when new software is tested.

The "System" page (Fig. 4) contains all the settings related to the VM hardware abstraction. If KVM is checked, KVM instead of QEMU is used and the menu for the selection of the guest machine architecture is disabled, as the only possible architecture is that of the host. The amount of RAM available must be specified in the "RAM" field. In the "Display options" section, it is possible to choose whether and how a virtual monitor should be generated. By default (no check box selected), a new window with a text terminal interface is created. Other options are "Disable graphical output", or "Start in VNC server", which creates a new VNC server process on a standard VNC display port, identified by the number specified in the text area. Among the other options in the same page, there is the possibility to synchronize the VM clock with that of the host, a feature that is very useful in dealing with the evaluation of time sensitive protocols, like TCP.

The "Network" page is used to add, remove, configure and interconnect the NIC interfaces. Finally, the "Customize Linux Boot" page is normally used only to debug a Linux kernel running on the guest.

2.3   VDE Bricks

VDE is an Ethernet compliant virtual network designed to interconnect VMs and real computers in the same Local Area Network (LAN) [8, 9]. It consists of many independent tools, such as switches, channel emulators, tap devices, tunnel servers and clients, etc., running on the host. It must be stressed that, like Qemu and KVM, VDE is a software package independent form Virtualbricks. Once VDE is installed on the host, Virtualbricks offers the user a GUI to manage VDE tools, seen as "VDE bricks", and

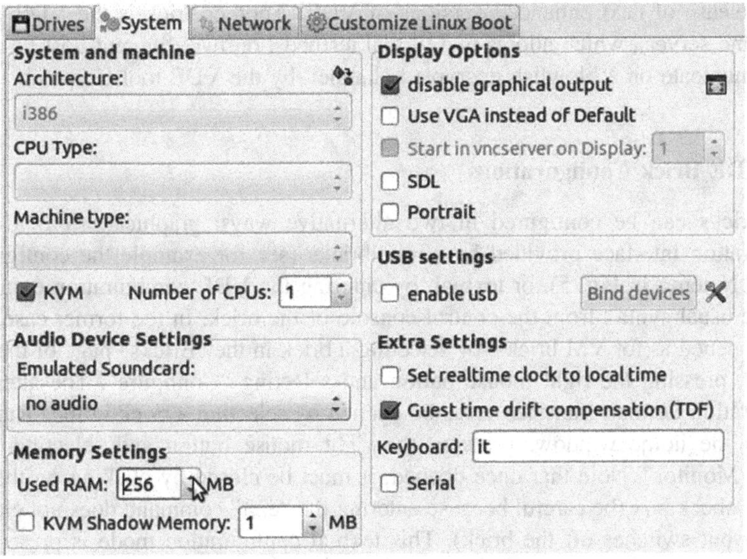

Drives System Network Customize Linux Boot

System and machine

Architecture:

i386

CPU Type:

Machine type:

☑ KVM   Number of CPUs: 1

Audio Device Settings

Emulated Soundcard:

no audio

Memory Settings

Used RAM: 256   MB

☐ KVM Shadow Memory: 1   MB

Display Options

☑ disable graphical output

☐ Use VGA instead of Default

☐ Start in vncserver on Display: 1

☐ SDL

☐ Portrait

USB settings

☐ enable usb   Bind devices

Extra Settings

☐ Set realtime clock to local time

☑ Guest time drift compensation (TDF)

Keyboard: it

☐ Serial

Fig. 4. The System configuration page of a VM brick

thus the possibility to interconnect VMs in many ways. This paves the way to the building of complex testbed layouts, which is what really differentiates Virtualbricks from all other VM management platforms. Of course, as in all virtual testbeds, there will be stringent limits to the maximum bandwidth of VDE tools supported, due to host hardware, CPU load and to the VDE software itself. However, as bandwidths of a few tenths of Mbit/s can be reached, there are plenty of possible applications, except the emulation of high-speed networks.

As in a real Ethernet, the most important tool is the VDE Switch, which emulates a real switch and is the most common way of interconnecting other VMs or VDE bricks. Moreover, by means of a VDE Tap brick, which provides the abstraction of an Ethernet device on the host machine, it is possible to connect the host to the switch, thus enabling the creation of a variety of hybrid testbeds: many VMs plus the host, distributed virtual testbeds, a virtual testbed linked to a real network through its host, etc. Alternatively, it is also possible to start a VDE Switch on the host independently from Virtualbricks and connect a Switchwrapper brick to it. This is the solution preferred by the authors to have remote access to the VMs by SSH (Secure SHell), which is very useful especially if the virtual testbed runs on a remote host. Another essential tool in networking evaluations is the VDE Wirefilter, usually inserted between two switches. Despite its name, it is a channel emulator. Like channel emulators on real machines, e.g. NistNet, Dummynet etc., it aims to reproduce the characteristics of a real wired or radio link, by making it possible to add packet delays, losses, bandwidth limitation, random duplication of packets, MTU restrictions and even the random flipping of single bits. It also supports asymmetrical channels. In Virtualbricks 1.0, however, we use an enhanced version of it, called Netemu, with increased reliability, not included in VDE (it can be downloaded by our web site). This is just a temporary solution in view of the release of next enhanced versions of VDE. Another brick is the VDE Tunnel (client and server), which allows two virtual testbeds, on two different host machines, to communicate on a blowfish encrypted channel, by the VDE tool Cryptcab.

2.4   VDE Brick Configurations

VDE bricks can be configured in two alternative ways: graphical, from a simple configuration interface provided by Virtualbricks (see for example the configuration page of Netemu in Fig. 5), or textual, by entering the VDE configuration commands with the usual syntax from the control console of the brick. In the former case, GUIs can be opened as for VM bricks, by selecting a brick in the "Bricks" page of the home window, pressing the right mouse button and selecting "Configure". The alternative configuration terminal can vice versa be opened by selecting a brick in the "Running" page of the home window, pressing the right mouse button and selecting "Open Control Monitor". Note that once opened, it must be closed by clicking on the close window check box (be careful because entering the "exit" command does not close the window but switches off the brick). This textual configuration mode is preserved in Virtualbricks to allow advanced users already familiar with VDE commands to continue to use VDE syntax, which is less intuitive but sometimes more powerful than GUIs.

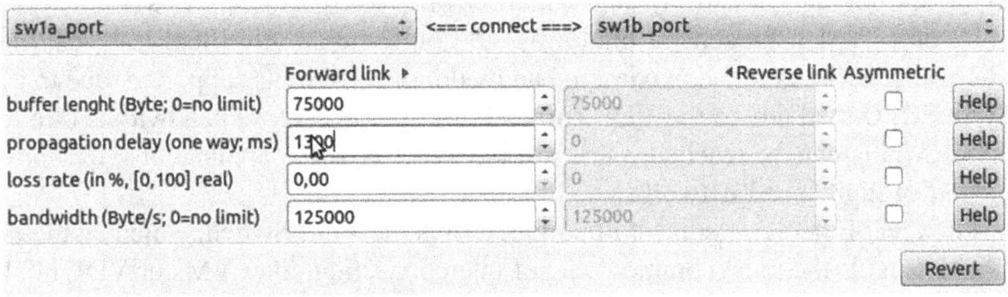

Fig. 5. The configuration page of Netemu brick.

3   Building a Virtual Testbed: DTN Satellite Communications in Space

One of the most important advantages of Virtualbricks over a real testbed is that it is possible to change the testbed layout simply by opening a different project. From the "File" menu it is possible to open an existing project file, create a new one, change the name of the current project, export the entire project (included VMs file systems) into just one file, or vice versa to import a project from a file. These features result into an extraordinary flexibility and easiness to use, for both research and education purposes. Since in virtual testbeds the available processing power of the host machine is a critical factor, as it limits performance, in Virtualbricks only one project can be opened at a time and only one Virtualbrick instance can be run on the same host.

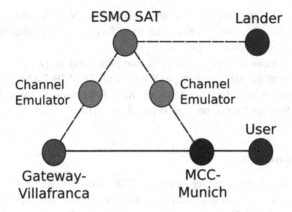

Fig. 6. The logical topology of the ESMO testbed.

Let us now introduce as an example the virtual testbed used in [17], to some extent inspired by the ESMO (European Student Moon Orbiter) ESA mission [20]. Its logical layout is shown in Fig. 6.

3.1 Brick Selection

To build it, we have to open a new project, which will call ESMO. Then we need to select the bricks. As the topology consists of five DTN nodes, we need five VM bricks. The fastest way to do this is to create a new VM, configure it, then clone it four times. Minor adjustments to the VMs, such as NIC connections, can be done after inserting the necessary VDE bricks into the project. As the two satellite links between the ESMO satellite, which orbits around the Moon, and the two ground stations (MCC and Gateway) are characterized by long propagation delays and possible random losses, their emulation requires the use of two intermediate Netemu bricks. All other links are

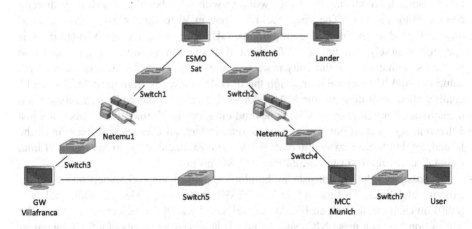

Fig. 7. The Virtualbricks topology of the ESMO testbed.

considered ideal. As direct connections are not possible, VDE Switches are necessary to connect VMs to each other, or a VM to a Netemu brick, thus leading to the Virtualbricks topology shown in Fig. 7.

The five VM bricks will have the same hardware and software configuration, as they are the clones of the same prototype: KVM, with 256 MB RAM, no CD, one hard disk, the same image including the OS and the DTN protocols to be tested, private cows and time alignment with the host (Figs. 3 and 4).

3.2 Brick Connections

In VDE the primary connection element is the VDE Switch, because it is the only brick that has the necessary sockets to connect the plugs of all the other bricks. We can start by connecting the VDE Netemu bricks (and other VDE bricks, if present) to the appropriate switches, by indicating the left and the right switch from the "Plugs" page of the Netemu configuration GUI (Fig. 5). Then, VMs must be connected to the switches. To this end, one NIC must first be created for each connection, through the "Network" page of the VM configuration GUI. Here three parameters must be set: the name of the switch to be connected to, the virtual NIC model and the MAC address. Regarding NICs models, note that the emulation of a real NIC is often partial. For example, the 10/100 Mbit/s NICs do not actually perform any bandwidth reduction, which can be misleading. The MAC address can be inserted either at random or manually. Parameters can of course be changed later, whenever necessary.

3.3 IP Address Assignment

Although in theory the IP address assignment in a virtual testbed is the same as in a real testbed, in practice there are some aspects that deserve to be discussed, to better highlight the scope of Virtualbricks.

In a real testbed the IP address of each machine is usually assigned (at least the first time) from a terminal (i.e. by a user working with a keyboard and a display directly connected to machine). The same could be done in Virtualbricks from VM consoles, once activated from the VM configuration GUI. In practice, when Virtualbricks is operated remotely through a VNC client this solution is impaired by translation problems of characters or difficulty in keeping the mouse control. An apparently simple solution could be to let the user assign the IP address directly from the VM "Network" configuration, as is done for the MAC address. However, it should be remembered that the software configuration of VMs is beyond the scope of Virtualbricks. This is not just a theoretical objection but also a practical one. In fact, all OSs allow the setting of the IP address, but the syntax is different. While the host machine has to run a GNU/Linux (this OS is required to run Virtualbricks), VMs do not.

A possible solution consists in building a control network independent of the experimental links. To this end, in our ESMO testbed we added an additional NIC (eth0) on each guest, with an IP address assigned by a DHCP server on the host. The connection between these NICs and the host is achieved by means of a Switchwrapper brick connected to a VDE Switch running on the host. This control network offers an easy access to all VMs by SSH as soon as they are created, from either the host or whatever node on Internet (provided that the host itself is on Internet, of course). Moreover, this control network is also very useful when experiments are carried out, as it provides the user with a parallel access system independent of the network under study, thus avoiding any possible interference.

4   Virtual vs. Real Testbed: Validation of Virtual Results

As Virtualbricks is a frontend for Qemu/KVM and VDE, the accuracy of its results actually depends on the reliability of these two components. An in-depth assessment of these was previously carried out by the authors in [21, 22] and would be beyond the scope of the present paper. However, for the sake of completeness, let us present a brief comparison between some results presented in [17], obtained with the Virtualbricks testbed described here, and those achievable by a real equivalent. A selection of Virtualbricks results from Fig. 7 of [17] is given in Fig. 8. The aim of the experiment was to assess the ability of ION CGR (Contact Graph Routing) [23], a DTN routing protocol designed by NASA for scheduled intermittent connectivity, to take the right decisions in the presence of parallel paths. Here from ESMO Sat to MCC, the routing alternative is either via the Gateway or directly to MCC. To this end, ten bundles [15] are first generated and taken into custody [14] on the Lander; when the first Lander-Sat contact starts, at 20 s, the first six are transferred to Sat and taken in custody; they are then delivered to MCC via Gateway when the Sat-GW contact opens, at 70 s; the other 4 bundles are transferred to Sat when the second Lander-Sat contact starts, at 100 s; then they are directly delivered to MCC when the Sat-MCC contact begins, at 150 s.

To replicate the test on a real testbed, we have set-up a real equivalent, consisting of 7 GNU/Linux machines (one for each VM, plus two for the channel emulators), running exactly the same code. Results are reported as x-crosses for comparison. The accuracy of the results achieved on the virtual testbed is evident, thus confirming once again that virtualization technologies can be used to carry out experiments on DTN satellite networking, where transmission rates are relatively low and in line with the present limits of virtualized links.

Of course, the higher the Tx rates and the higher the number of VMs, the higher the chances of reaching the limits of virtualization. For this reason, we do not suggest using virtualization technologies for high speed networks; we do however deem virtualization a perfect match for satellite communication in general, because of low Tx rates, and with DTN in particular, because of both low Tx rates and link intermittency. The unavailability of many links for relatively long periods, typical of most DTN environments, including LEO sat communications and deep space networks, results in intermittent use of VMs, i.e. in a lower computational load for the host machine. In other words, the processing power of the host CPU is shared only by the fraction of active VMs, which can be small even in a large testbed.

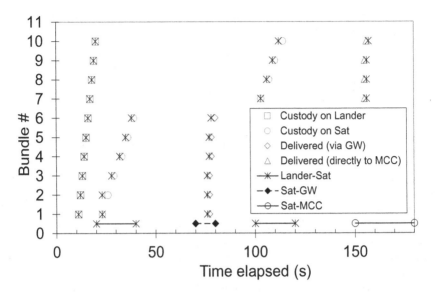

Fig. 8. Bundle transfer from Lander to User. Markers: Virtualbricks; x-crosses: real testbed; segments at the bottom: contact windows.

5   Research and Education with Virtualbricks

The testbed shown in the previous section was successfully used with minimal modifications in further research on Moon communications [24] and in LAB activities of the course [16]. Other Virtualbricks testbeds were used to develop and test DTNperf_3 [25], and more recently to carry out joint research on DTN routing with SPICE center of Democritus University of Trace (Greece). This clearly proved the potential of the tool for both research and education.

For fairness, we also remember the limits of virtualization. They consist in the unfeasibility of carrying out high-speed tests and in software maintenance, which is only apparently the same as in a real testbed. In fact, the possibility to "buy" unlimited VMs at no cost, may lead to the unnecessary proliferation of testbeds, whose software maintenance requires the same effort of their real equivalents (e.g. for updating OS or applications). The informed user, aware of this risk, will easily avoid the problem.

For the reader convenience, let us conclude this section by listing the advantages provided by Virtualbricks in both research and education.

5.1   Research

- Reduced TCO (total cost of ownership): no need to buy dedicated hardware.
- Use of real protocol stacks, by contrast to network simulators.
- Very good performance by using KVM for VMs, by contrast to emulators; alternatively, high flexibility in the CPU architecture choice by using Qemu.

- Perfect results reproducibility: as a Virtualbricks testbed is software defined, two independent research team can actually work on the very same testbed and found the very same results.
- Increased productivity: no more one real testbed to be time-shared among many researchers of the same team, but independent testbeds to be used in parallel.

5.2 Education

- Reduced cost: no need of dedicated LABs.
- Increased teacher productivity: no need to maintain/update/change the configuration of a real testbed; no need to organize testbed remote access and sharing among the students.
- Easy installation: Virtualbricks is in Debian and can be installed with the usual commands (apt-get install Virtualbricks). Note, however, that at present ver.1.0, just released, must be downloaded from [12].
- No need to set-up testbeds: once installed Virtualbricks, it is easy to import pre-configured testbeds provided by the teacher.
- No more one testbed fits all: a set of pre-configured testbeds (e.g. one different testbed for each LAB activity) can be downloaded by students from a course web site and easily imported into Virtualbricks.
- Increased student productivity and freedom; students can focus their attention on the aim of the LAB activity, without being distracted by the many practical problems related to the remote access and time-sharing of a physical testbed; the presence of a testbed in their own PC allows students to replicate LAB activities at home, or to carry out their own experiments, at their will.

6   Conclusions

In the paper the main features of Virtualbricks have been presented. This virtualization solution for Linux differs from others because of the support not only of VMs (Qemu and KVM), but also of VDE tools, which makes Virtualbricks particularly useful for designing and managing testbeds consisting of multiple interconnected VMs. For this reason, in the description of Virtualbricks the virtual testbed used by the authors in recent research on DTN satellite communications has been considered, as a real application example. The comparison of Virtualbricks results with those achievable with an equivalent real testbed, has shown an excellent level of accuracy, thus confirming the suitability of the virtualization approach for both DTN satellite communication research and education.

References

1. VMware. http://www.vmware.com
2. SCVMM.   http://www.microsoft.com/en-us/server-cloud/system-center/virtual-machine-manager.aspx
3. Paragon VM. http://www.paragon-software.com/home/vm-professional/
4. SolarWind. http://www.solarwinds.com/
5. KVM. http://www.linux-kvm.org/page/Main_Page
6. Loddo, J., Saiu, L.: Status report: Marionnet - how to implement a virtual network laboratory in six months and be happy. In: Proceedings of the ACM SIGPLAN Workshop on ML, pp. 59–70. ACM Press, New York (2007)
7. Loddo, J., Saiu, L.: Marionnet: a virtual network laboratory and simulation tool. In: SimulationWorks, Marseille, France (2008)
8. Davoli, R.: VDE: virtual distributed ethernet. In: Proceedings of ICST/Create-Net Tridentcom 2005, Trento, Italy, pp. 213–220, May 2005
9. VDE. http://vde.sourceforge.net/
10. UML. http://user-mode-linux.sourceforge.net/
11. Mininet. https://github.com/mininet/mininet/wiki/Introduction-to-Mininet
12. Virtualbricks. https://launchpad.net/virtualbrick
13. Qemu. http://wiki.qemu.org/Main_Page
14. Caini, C., Cruickshank, H., Farrell, S., Marchese, M.: Delay- and disruption-tolerant networking (DTN): an alternative solution for future satellite networking applications. Proc. IEEE 99(11), 1980–1997 (2011)
15. Cerf, V., Hooke, A., Torgerson, L., Durst, R., Scott, K., Fall, K., Weiss, H.: Delay-Tolerant Networking Architecture. Internet RFC 4838, April 2007
16. TLC Master course on Architectures and Protocols for Space Networks. http://www.engineeringarchitecture.unibo.it/en/programmes/course-unit-catalogue/course-unit/2013/386378
17. Caini, C., Fiore, V.: Moon to Earth DTN communications through lunar relay satellites. In: Proceedings of ASMS 2012, Baiona, Spain, pp. 89–95, September 2012
18. KSM. http://www.linux-kvm.org/page/KSM
19. Binary Translation. http://en.wikipedia.org/wiki/Binary_translation
20. ESMO. http://www.esa.int/esaMI/Education/SEML0MPR4CF_0.html
21. Caini, C., Davoli, R., Firrincieli, R., Lacamera, D.: Virtual integrated TCP testbed (VITT). In: Proceedings of ICST/Create-Net Tridentcom 2008, Innsbruck, Austria, pp. 1–6, March 2008
22. Caini, C., Firrincieli, R., Lacamera, D., Livini, M.: Virtualization technologies for DTN testbeds. In: Proceedings of PSATS 2010, Rome, Italy, pp. 272–283, February 2010
23. ION code. http://sourceforge.net/projects/ion-dtn/
24. Apollonio, P., Caini, C., Fiore, V.: From the far side of the Moon: DTN communications via lunar satellites. China Commun. 10(10), 12–25 (2013)
25. Caini, C., d'Amico, A., Rodolfi, M.: DTNperf_3: a further enhanced tool for delay-/disruption- tolerant networking performance evaluation. In: Proceedings of IEEE Globecom 2013, Atlanta, USA, pp. 3009–3015, December 2013

Research Challenges in Nanosatellite-DTN Networks

Marco Cello$^{(\boxtimes)}$, Mario Marchese, and Fabio Patrone

University of Genoa, Genoa, GE, Italy
{marco.cello,mario.marchese}@unige.it, f.patrone@edu.unige.it
http://www.scnl.dist.unige.it/

Abstract. Current approaches based on classical satellite communications, aimed at bringing Internet connectivity to remote and underdeveloped areas, are too expensive and impractical. Nanosatellites architectures with DTN protocol have been proposed as a cost-effective solution to extend the network access in rural and remote areas. In order to guarantee a good service and a large coverage in rural areas, it is necessary to deploy a good number of nanosatellites; consequentially, for reliability and load balancing purposes, is also needed a large number of ground stations (or hot spots) connected on the Internet. During a data connection, a server on the Internet that wants to reply to the user on rural area, has many hot spot alternatives to whom it can deliver data. Different hot spots can send data to final destination with different delivery delay depending on the number, position and buffer occupancy of satellites with which it comes into contact. The problem of choosing the optimal hot spot becomes important because a wrong choice could lead a high delivery delay.

Keywords: Nanosatellite network · Delay tolerant network architecture · Congestion control · Next-hop selection

1 Introduction

Despite the worldwide demand of ICT services and the continuous increment of the number of developing countries, currently, only about 40 % of the world population has access to Internet. One of the reasons is that a large amount of people still lives in underdeveloped countries or in remote areas which do not possess ICT infrastructure. The costs needed to connect these areas using cables and common infrastructures are prohibitive compared with the yielded benefits. Satellite communications provide a less expensive way to provide Internet access in these areas. However, current satellite technologies require high costs in the construction, launch and maintenance. Nanosatellites [1] have been recently proposes as a cost-effective solution to extend the network access in rural and remote areas. CubeSat [2], a kind of nanosatellite, is fabricated and launched into low-earth orbit using 0.1 % of the cost of a classical LEO communication satellite. Rural and/or disconnected area will be connected through

© ICST Institute for Computer Sciences, Social Informatics and Telecommunications Engineering 2016
I. Bisio (Ed.): PSATS 2014, LNICST 148, pp. 89–93, 2016.
DOI: 10.1007/978-3-319-47081-8_8

a local gateway (*cold spots*) that will communicate in an opportunistic fashion with the nanosatellite constellation using the Delay Tolerant Networking (DTN) paradigm. Nanosatellites will carry the data and will send them to the gateways connected to the Internet. On the return path, the central node of the constellation will communicate with servers on the Internet and with the nanosatellite through deployed hot spots that will deliver the data to the rural area.

2   Related Works

The problem to connect remote areas to the Internet is not a recent challenge. [3] proposes to establish a communication with Internet for the nomadic Saami population who lives in remote areas in Swedish Lapland. The solution uses DTN mobile devices and a series of fixed and mobile relay nodes. In [4] is described DakNet, an ad-hoc wireless network that provides asynchronous connectivity. DakNet is based on rural kiosks to deliver information to users and portable storage devices called Mobile Access Points (MAPs) mounted on a bus, a motorcycle or even a bicycle, which transport data among kiosks and Internet gateways. A similar architecture is described in [5]. The architecture in [6] is a multi-hop mesh network composed of long-distance 802.11 links with high gain directional antennas.

All the described architectures offer valid and inexpensive solutions (e.g. with an investment of $15 million, DakNet could equip 50 000 rural buses in India), but suffer of severe performance limits and insecurities due to the massive use of ground facilities.

To bypass these drawbacks, satellite networks have been proposed as solution. Iridium [7], Globalstar [8] and Orbcomm [9] are Low Earth Orbit (LEO) satellite constellations that provide satellite phone and low-speed data communications. Inmarsat [10] is a Geostationary Earth Orbit (GEO) satellite constellation that provides voice and data communication services. Nevertheless, these solutions are very expensive due to the production and launch costs. Other solutions involve the use of a network of balloons traveling at an altitude of about 20 kms (Google's Project Loon) [11], and the use of drones in the new Facebook project called Internet.org [12].

A recent solution [1] is represented by the joint use of nanosatellites and DTN paradigm. Nanosatellite is an interesting solution aimed of avoiding the drawbacks of the use of an all-terrestrial network and to reduce the implementation costs of GEO and LEO satellite networks. CubeSat [13] is a nanosatellite: it is a 10 cm cube with a mass up to 1.33 kg. The main advantage of CubeSats is the reduced cost: the estimated assembly cost per satellite is from $50 000 to $100 000, while the estimated launch cost per group of three Cubesats is about $200 000. The total cost of a possible CubeSat network composed of 150 nanosatellites and 3 000 base stations is about $33 million with a lifetime of 5 years.

The DTN paradigm, on the other hand, allows supporting end-to-end data exchange between network nodes even when network paths are concatenations of time-disjoint transient communication links. The DTN architecture [14] is based on the introduction of an overlay layer above transport layer protocol which allows to handle delays and disruptions at each hop in a path between a sender and a receiver [15]. The principal implementation of DTN is the Bundle Protocol (BP) [16] whose PDU is the bundle.

3 Motivations and Use Case Scenario

The access network on rural areas we envision is composed of a constellation of simple, inexpensive nanosatellites that communicate with ground stations through the DTN paradigm. Figure 1 shows a nanosatellites/DTN network scenario: in a rural area, a group of users or nodes $S_1, \ldots, S_N$ is connected with the node CS_1. Nodes CS_1 and CS_2, referred in the following as cold spot (CS) are located in remote areas and act as Internet gateway for users. They transmit and receive data with nanosatellites SAT_1, SAT_2, SAT_3. Node D is the destination node (e.g. a mail server on the Internet). Node C is the control node of the nanosatellite constellation: it contains all the information necessary to manage the network and takes the decisions to improve the performances. Finally, nodes HS_1 and HS_2, referred in the following as hot spot (HS), are connected to Internet and able to exchange data with satellites.

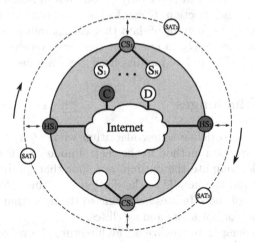

Fig. 1. Nanosatellite network scenario.

Referring to Fig. 1, we present a use case scenario in which a user S_1 located in a rural area wants to access a web page located in a web server D on Internet. User S_1 sends a DNS request to its default gateway, the cold spot CS_1, which is also its DNS server. CS_1 replies to the DNS request pretending to be the web server. S_1 establishes a TCP connection with CS_1 and send it the HTTP GET request. CS_1 replies to S_1 that the HTTP request has been taken in charge and it will reply as soon as it gets the web page from the web server. CS_1 encapsulates the HTTP GET message in a bundle destined to central node C on Internet and uploads it on the first satellite it comes in contact with (e.g. SAT_1). The bundle is carried by satellite SAT_1 until it comes in contact with hot spot HS_2. HS_2 receives the bundle and sends it to central node C by using TCP/IP standard protocols. Central node de-encapsulates the bundle to obtain the HTTP GET message and pretending to be the user, starts a TCP connection with the web server D. After the reception of the web page, C creates one or more bundles that are forwarded to the selected HS, then to the first satellite that can upload them, and finally delivered to CS_1. CS_1 de-encapsulates the bundle and send back the web page to S_1 by using the same TCP connection of the initial request.

The choice of the hot spot, as said in the introduction, has a direct impact on the delivery time, which should be minimized. This choice can be static (e.g. C always forwards all messages destined to a certain CS to the same HS) or dynamic. Referring to Fig. 1, we suppose that 100 bundles are destined to CS_1 and others 100 bundles are destined to CS_2. Because of the limited communications performances between hot spot and nanosatellites, only a given amount of data can be uploaded by the HS to the satellite during each contact: in this example we suppose that only 10 bundles can be uploaded. With a static choice, C forwards all 200 bundles to HS_1. 10 satellite contacts are necessary to deliver 100 bundles to CS_1 and other 10 to deliver 100 bundles to CS_2 because each satellite in each orbit time can carry only 10 bundles uploaded by HS_1. Alternatively, with a dynamic selection, C can forward 100 bundles to HS_1, and 100 to HS_2 thus doubling the amount of data that each satellite can upload during each orbit. For example, SAT_3 in one orbit time may receive 10 bundles from HS_2 and destined to CS_2 and 10 bundles from HS_1 and destined to CS_1.

4   Research Challenges

The first challenge is to define a new algorithm whose purpose is to realize a dynamic hot spot selection method in the central node C. For each bundle, the central node should compute the optimal hot spot that minimize the delivery time necessary to send the bundle to the destination using information such as the current position of the satellites belonging to the orbit that it manages, and the buffer occupancy of hot spots and satellites.

The second challenge is to realize an architecture (based on [17,18]) which ensures a transparent communication between endpoints: users on rural areas make use of standard devices with TCP/IP protocol stack. No no-standard protocols or protocol modifications on the users' devices are allowed. In the same way, server nodes on Internet must use standard protocols. Differently from the literature about DTN, that assume bundle protocol installed on endpoints, for transparent purposes we want to be installed only on cold spots, hot spots, and central node. To do this we need to design a novel architecture able to guarantee on one hand, TCP/IP protocol communications among endpoints and on the other hand bundle protocol and satellite-specific transport protocol for the link section. This architecture is illustrated in Fig. 2.

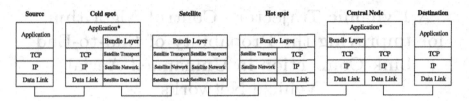

Source	Cold spot		Satellite		Hot spot		Central Node		Destination
Application	Application*						Application*		Application
		Bundle Layer	Bundle Layer		Bundle Layer		Bundle Layer		
TCP	TCP	Satellite Transport	Satellite Transport	Satellite Transport	Satellite Transport	TCP	TCP	TCP	TCP
IP	IP	Satellite Network	Satellite Network	Satellite Network	Satellite Network	IP	IP	IP	IP
Data Link	Data Link	Satellite Data Link	Satellite Data Link	Satellite Data Link	Satellite Data Link	Data Link	Data Link	Data Link	Data Link

Fig. 2. Network Architecture

References

1. Burleigh, S.: Nanosatellites for universal network access. In: Proceedings of the 2013 ACM MobiCom Workshop on Lowest Cost Denominator Networking for Universal Access. ACM (2013)
2. Heidt, H., Puig-Suari, J., Moore, A., Nakasuka, S., Twiggs, R.: CubeSat: a new generation of picosatellite for education and industry low-cost space experimentation (2000)
3. Doria, A., Uden, M., Pandey, D.: Providing connectivity to the saami nomadic community. Generations **1**.2, 3 (2009)
4. Pentland, A., Fletcher, R., Hasson, A.: Daknet: Rethinking connectivity in developing nations. Computer **37**(1), 78–83 (2004)
5. Seth, A., Kroeker, D., Zaharia, M., Guo, S., Keshav, S.: Low-cost communication for rural internet kiosks using mechanical backhaul. In: Proceedings of the 12th Annual International Conference on Mobile Computing, Networking: Observation of Strains, pp. 334–345. ACM (2006)
6. Raman, B., Chebrolu, K.: Experiences in using WiFi for rural internet in India. IEEE Commun. Mag. **45**(1), 104–110 (2007)
7. Iridium Global Network. http://www.iridium.com/About/IridiumGlobalNetwork.aspx
8. Globalstar Network. http://eu.globalstar.com/en/index.php?cid=3300
9. Orbcomm Networks. http://www.orbcomm.com/networks
10. Inmarsat Satellites. http://www.inmarsat.com/about-us/our-satellites
11. Google Project Loon. http://www.google.com/loon/
12. Facebook and Partner's Project Internet.org. http://internet.org/
13. Munakata, R.: Cubesat design specification rev. 12. The CubeSat Program, California Polytechnic State University **1**, (2009)
14. Cerf, V., Burleigh, S., Hooke, A., Torgerson, L., Durst, R., Scott, K., Fall, K., Weiss, H.: Delay-tolerant networking architecture. RFC4838, April 2007
15. Caini, C., Cruickshank, H., Farrell, S., Marchese, M.: Delay-and disruption-tolerant networking (DTN): an alternative solution for future satellite networking applications. Proc. IEEE **99**(11), 1980–1997 (2011)
16. Burleigh, S., Scott, K.: Bundle protocol specification. IETF Request for Comments RFC 5050 (2007)
17. Guo, S., Falaki, M.H., Oliver, E.A., Rahman, U.S., Seth, A., Zaharia, M.A., Keshav, S.: Very low-cost internet access using KioskNet. ACM SIGCOMM Comput. Commun. Rev. **37**(5), 95–100 (2007)
18. Scott, K.: Disruption tolerant networking proxies for on-the-move tactical networks. In: Military Communications Conference, 2005, MILCOM 2005. IEEE (2005)

A Dynamic Trajectory Control Algorithm for Improving the Probability of End-to-End Link Connection in Unmanned Aerial Vehicle Networks

Daisuke Takaishi[1,2(✉)], Hiroki Nishiyama[1,2], Nei Kato[1,2], and Ryu Miura[1,2]

[1] Graduate School of Information Sciences, Tohoku University, Sendai, Japan
[2] Wireless Network Research Institute, National Institute of Information
and Communications Technology, Koganei, Tokyo, Japan
{takaishi,bigtree,kato}@it.ecei.tohoku.ac.jp, ryu@nict.go.jp

Abstract. *Recently,* the Unmanned Aircraft Systems (UASs) have attracted great attention to provide various services. However, the Unmanned Aeria Vehicle (UAV) network which is constructed with multiple UAVs is prone to frequent disconnection. This is why the UAV-to-UAV links are constructed with two UAVs with high mobility. In such a disconnected network, ground-nodes cannot communicate with other ground-nodes with End-to-End link and the communication failure. Because the UAVs fly along with a commanded trajectory, the trajectories are the most important to decide UAV network performance. In this paper, we propose a effective UAVs' trajectory decision scheme.

Keywords: Unmanned Aircraft System (UAS) · Unmanned Aerial Vehicle (UAV) · End-to-End link connection

1 Introduction

Recent advances in wireless communication technologies and autonomous control technologies have made the Unmanned Aircraft System (UAS) applications feasible. UAS is a system made up of multiple Unmanned Aerial Vehicles (UAVs), which are small aircraft vehicles equipped with sensors, video camera, and wireless communication modules. UAV flies over the ground with propeller empowered by equipped battery, and use equipment to gather the information. Gathered information by UAVs are transmitted to ground-nodes (e.g. mobile phones, Access Points (APs), sensors and so forth) by using wireless communication modules. Generally, these UAVs are controlled by a control station located on the ground. UAVs receive the trajectory command from the remote control station, and travel along with transmitted trajectory. These UAVs can be classified into fixed-wing UAVs and rotor-propelled UAVs. Fixed-wing UAVs can fly with a higher speed than rotor-propelled UAVs. Moreover, fixed-wing UAVs can fly longer distance than rotor-propelled UAVs but cannot stay stationary at a

© ICST Institute for Computer Sciences, Social Informatics and Telecommunications Engineering 2016
I. Bisio (Ed.): PSATS 2014, LNICST 148, pp. 94–105, 2016.
DOI: 10.1007/978-3-319-47081-8_9

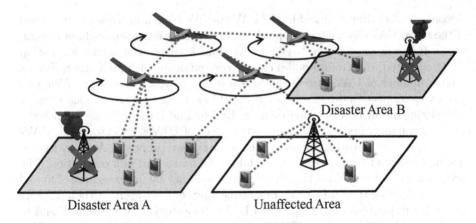

Fig. 1. Multi-hop data communication employing UAV network.

location. Therefore, it is clear that fixed-wing UAVs can better to provide the applications in large areas (e.g. urban area, mountains, islands and so forth) with its high mobility. On the other hand rotor-propelled UAVs can provide the applications such as fixed point observations by hovering objectives. The applications made possible by UAVs include scouting hazardous areas [1,2], collect data from mobile sensors [3], environmental observation [4–6], and so forth. Additionally, the UAVs trajectories can be dynamically changed in real-time by the control station to achieve these applications' objectives. Hereafter, we refer to a fixed-winged UAV as a UAV for brevity because our objective is to provide the services in wide area.

Relaying the data from ground-nodes to other ground-nodes is one of the anticipated UAS applications. This kind of application is especially useful when deployed over the disaster areas where conventional networks (e.g., antennas, ground base stations, network cables, etc.) are damaged and stopped. In such disaster area, conventional network infrastructures loses ability to provide the network connectivity. UAV network, which is constructed with multiple UAVs, can provide the connectivity to the ground-nodes which is distributed on those areas by using equipped communication module. The transmitted data from ground-nodes is received by flying UAV over the ground-nodes. The received data are transmitted to the destination ground-nodes in a multi-hop fashion by employing the UAV network. An overview of network construction is shown in Fig. 1. In Fig. 1, UAV networks can relay the data among the areas by connecting the wireless communication link to the each ground-node. Generated data in disaster areas are transmitted to the base station in the non-affected area.

However, UAV network's relay communication is not always successful because of the distance limitation of wireless communication. In the case that the distance between communicating nodes are larger than the limitation of wireless communication, the communication fails. Therefore, the ground-nodes cannot send the data when there are no UAVs inside of communication range. The link disconnections is more critical in the UAV-to-UAV communication because both of these two UAVs fly with high speed and easily move outside of communication range. If one or more of the UAV-to-UAV links between source and destination is disconnected, the ground-nodes cannot communicate with each other. Even if a large number of UAVs are deployed, we still need to consider the UAVs' trajectory to connect End-to-End link. If the UAVs' trajectory are decided without considering about network environment, End-to-End links are not established.

In this paper, firstly, we calculate the effect of UAV's trajectories on UAV-to-UAV links connection. Based on the analysis, we propose UAV's trajectory decision scheme to enhance the probability of End-to-End link connection. The proposal scheme calculates the each nodes' trajectory by using volume of flowed packet to provide the End-to-End link connection to many users. Although, there are so many parameter (e.g., shape of UAV's trajectory, altitude, speed, and so forth), we suppose that all of UAV have circular trajectories. The center position vector and the radius of circular trajectory can be changed by the control station. This is a reasonable assumption that UAV need to cover users who are around damaged base station while operation in the disaster situation.

The remainder of the paper is organized as follows. Section 2 reviews some related works and presents our research motivation. In Sect. 3, we show our ground node aware clustering algorithm. Performance evaluation is presented in Sect. 4. Finally, Sect. 5 concludes the paper.

2   Related Works and Our Motivations

The network construction with vehicles studied in some areas [7,8]. Mobile sink is the one of the network construction by using a vehicle. In the Mobile sink scheme, movable sink (e.g., vehicle, Unmanned Aerial Vehicle and so on) patrols the Wireless Sensor Networks (WSNs). As the sink node moves around the network area, the sensor nodes send data to the sink node when the sink node comes in their proximity. Thus, energy consumption can be decreased by reducing the amount of relays in the WSN. However, mobile sink make the big delay because the mobile sink moves to proximity of sensor nodes. In [9,10], the authors proposed the data aggregation method within limited period or limited buffer. The minimizing sum of required energy for data aggregation with a mobile sink are proposed in [11].

In [12], the authors proposed a Message Ferrying (MF) scheme. Message Ferry (MF) scheme is a approach for routing in disconnected ad hoc networks. It address the disconnection problem by introducing MF's mobility. In the MF scheme, the some rendezvous points are calculated beforehand to connect the all of disconnected ad hoc networks. MF schemes are resemble to mobile sink schemes and UAV networks. In [13], the author propose the hierarchical structure of message ferry data transmission to improve the network capacity. Although the MF scheme connect between disconnected ad hoc networks, these researches do not consider the End-to-End link connection. All of the received data are carried with MF's mobility.

The UAVs' trajectory decision scheme was proposed in [14, 15]. In [14], the authors proposed the real-time environment sensing scheme with multiple UAVs. The proposed scheme is a role based trajectory decision scheme and effective for sensing all field. This scheme, however, do not consider the frequent sensing. Moreover, the destination gathered by UAVs is a fixed.

In this paper, we present how to decide the UAVs' trajectory to provide the End-to-End connectivity to ground-users by using multiple UAVs. Based on a communication performance analysis between UAVs, we propose the simple UAV's trajectory decision scheme.

3   UAV's Trajectory Decision Scheme

UAVs' trajectory is one of the most important factors that decide the probability of End-to-End link connection in UAV networks. In this section, we first show how the UAV-to-UAV link performance affected by the UAVs' trajectory. Based on the UAV-to-UAV link performance analysis affected by UAVs trajectory, we propose a UAVs trajectory decision scheme to improve the probability of End-to-End link connection.

3.1   Network Model

In this paper, we consider a network that consists of a ground node (e.g., mobile sensors, mobile phones, Access Points (APs)) spreads within a limited field and UAVs are deployed over the field. The Wi-Fi technology can be easily deployed on UAVs and ground nodes. Regardless of being a ground nodes or UAVs, a node has a Wi-Fi's limited communication range r, and communication is always successful if it is conducted within r.

The ground-nodes are densely distributed in some areas, which iclude refuge sites in disaster struck areas. These ground nodes are supposed to transmit the data to other grounds nodes in refuge sites. Generally, these nodes communicate with the base station to transmit the data to the destination ground-nodes. And the base station transfers the data to base station by employing wired cables. However, in some networks such as those deployed in disaster areas, islands and so forth, base stations are not always connected with each other. Due to physical factor, ground-nodes cannot communicate with the destination. If the End-to-End link between ground-nodes and the destination is disconnected, communication fails.

UAV networks, which consist of several UAVs deployed over the field. These UAVs travel along a circular trajectory to provide network service to a certain area. By communicating with each other, UAVs can transfer the data in a multi-hop fashion. All of the UAVs' trajectories (which consist of the center position of trajectory, and the radius of trajectory) are determined and controlled by a control station. A control station deployed on the field, communicates with all of the UAVs and transmit the UAVs' position vectors, and receive data directed to UAV in real-time. UAVs control their own trajectories by comparing with information from GPS and the information sent from the control station. UAV fly along the commanded trajectory paths. The data from the ground-nodes are relayed to the destination ground node through multiple UAVs.

3.2 UAV-to-UAV Link Performance Analysis

In the UAV-to-UAV link composed of two UAVs having circular trajectory, the UAV link performance is affected by the center position vector of the circle and radius of the circle. When the distance between two UAVs is smaller than communication range, r, a communication link is established and the communication link is disconnected when the distance is larger than r. And also the phases of circle is one of the determinants of the link performance. However, we deal with the phases as the means to get to know how trajectory's shape affects the UAV network's performance. Figure 2 shows the UAV-to-UAV link factors. UAV_i has the circle trajectory with the radius of R_i and center position vector of $X_i(x_i, y_i)$. Each UAV has a limited communication range, r, and communication is successful if it is executed within r.

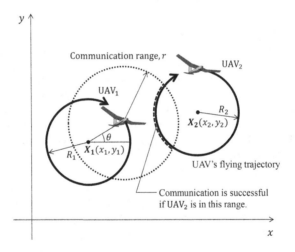

Fig. 2. The parameters of UAV-to-UAV link.

As shown in Fig. 2, when UAV_i is on trajectory with phase of θ, the successful communication probability is the ratio of length of lapped circle to length of circumference, $p(\theta)$, is shown as follows,

$$p(\theta) = \frac{1}{\pi} \arccos \left(\frac{R_2^2 + d^2 - r^2}{2 R_2 d} \right), \tag{1}$$

where d is the distance between UAV_1 and center of UAV_2s circular trajectory,

$$d = \sqrt{(x_2 - x_1 + R_1 \cos \theta)^2 + (y_2 - y_1 + R_1 \sin \theta)^2} \tag{2}$$

Therefore the average successful communication probability, $P_{1,2}$ (x_1, x_2, R_1, R_2), is shown as follows,

$$P_{1,2}(x_1, x_2, R_1, R_2,) = \oint_{UAV_1} p(\theta). \qquad (3)$$

According to 3, the radius of trajectory, R, and the center position vector of the circular trajectory, $\mathbf{X}$, affect the network performance. $P_{1,2}$ decrease with $\|\mathbf{X_2} - \mathbf{X_1}\|$. And $P_{1,2}$ decrease with R_i and R_j in some $\|\mathbf{X_2} - \mathbf{X_1}\|$.

3.3 Proposed UAVs' Trajectory Decision Scheme

In this subsection, we propose a effective UAVs' trajectory decision scheme. Our objective is to improve End-to-End connection probability. The proposed scheme dynamically and recursively changes the UAVs' trajectory based on the UAV-to-UAV link performance analysis.

In the assumed network environment, a control station controls the trajectory path of all of the UAVs by using allocated frequency bands for controlling the UAV. The allocated frequency bands is to communicate between control station and UAVs. General data transmissions on this bands are prohibited. Such remote trajectory controls are executed in real-time. According to the UAV-to-UAV link performance analysis, the center position vectors and the radius affect the UAV-to-UAV link's connection probability. UAV-to-UAV link's average connection probability can be increased by decreasing the radius of the circular trajectory and decreasing distance between two UAVs trajectories. The average link connection probability is not always a strict monotonic function of trajectory's parameters which include distance between UAV's trajectory and the radius of circular trajectory. However, we address the radius and the distance between UAV's trajectory while the average link probability is monotonically function, which means algorithm will stop when the average connection probability becomes a non monotonic function. An overview of our algorithm is shown in Algorithm 1.

At first, control station calculates the expected amount of successful communication in each link. Average link connection probability is calculated by 3 based on the UAV's trajectory information. After the calculation of expected number of successful communication, the algorithm selects the bottleneck link which is indicated by the link with the highest m_l.

In the selected link that has the highest value of m_l, our proposed algorithm selects the one of the UAV$_i$ that has the bottleneck link. According to the link performance analysis, link's average successful communication probability can be controlled by changing UAV's trajectories. Concretely speaking, the average probability of successful End-to-End link connection increases by shortening the distance between two UAVs' center position vector of circular trajectory. And the shortening the radius of circular trajectory is also effective solution to improve the link's communication probability. Therefore, by shortening the distance between two UAVs or the radius, the average communication speed is increased. Then the successful End-to-End communication probability is also increases.

Algorithm 1. Proposed clustering algorithm

Set initial UAVs trajectory
while $|m_i - m_j| \leq \epsilon$ **do**
   Calculate the **m**.
   /* Phase 1, Update the center position vector of UAS's trajectory */
   Select a bottleneck link, l.
   Select a UAV$_i$ and UAV$_j$ which compose the bottleneck link l
   Move the UAV$_i$ and UAV$_j$ to reduce the distance, $||\mathbf{X_i} - \mathbf{X_j}||$
   /* Phase 2, Update the radius of UAV's trajectory */
   Check the coverage area
   if All of network field is covered **then**
      Apply calculated UAV's trajectories
      $R_i' = R_i - \Delta$
      $R_j' = R_j - \Delta$
   else
      $R_k = R_k + \Delta$ for all UAV$_k$
      Apply calculated UAV's trajectories
   end if
end while

However, we also need to consider the coverage area of UAV network in addition to communication probability. If the UAVs' trajectories are updated with consideration about the communication probability and without consideration about coverage area, some ground-nodes may become not able to connect to UAV. If the ground-nodes are outside of UAVs' communication range, the nodes' data are no longer to reach destination. To decide the UAVs' trajectory with having high communication probability and making sure the UAVs cover all of the field, our proposed algorithm checks the coverage area by using some existing schemes [16]. Only when the calculated trajectory cover all of the network field, the trajectories are applied. In case that the UAV network does not cover all of the network field, the algorithm enlarge the radius of all UAVs trajectories. Then, each UAV changes the own trajectory to received one.

4   Performance Analysis

In this section, we measure the performance of the UAV network and evaluate the performance of the proposed UAV trajectory decision scheme through extensive computer simulations. The simulation scenario was configured with the parameters summarized in Table 1. The nodes (e.g. UAVs and ground-nodes) use 2.4 GHz Wi-Fi band to connect with each other without additional base stations. We set Wi-Fi communication range, r, as 150 m and communication is successful if it is conducted within r. We assume that ground-nodes are distributed in some area such as refuges, schools, studiums, and so forth.

UAV networks are constructed over the ground to provide network connectivity to all ground-users. These UAVs have circular trajectory and each trajectory can be controlled by the control station.

Table 1. Environment of experiment

Number of users	100
User distribution	Even
	Gaussian Mixture
Number of UAVs, N	3–10
Speed of UAVs, v	40 km/h, 80 km/h
Communication range, r	150 m
Length of one side of field	1000 m

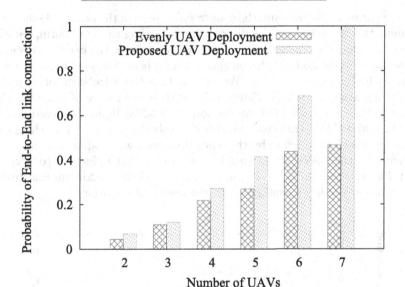

Fig. 3. Effect of the UAV's trajectory decision scheme.

We compare our UAV's trajectory decision scheme with even UAV deployment. The even UAV deployment is one where all of the UAVs have the same radius and the position vector of circular trajectories are uniformly deployed. On the other hand, the proposed scheme has the initial placement which is decided by even UAV deployment. Then proposed algorithm gradually change the UAV's deployment with UAV's speed.

4.1 End-to-End Link Connection Probability

In this experiment, we measure the End-to-End link connection probability from ground user to another ground user to evaluate our proposed algorithm in comparison to other trajectory decision schemes. Figure 3 shows average End-to-End link connection probability. As the graph shows the proposed algorithm can achieve higher End-to-End probability compared to uniform UAV deployment.

Since the proposed UAVs' trajectory decision scheme dynamically changes the trajectory to improve link connection probability based on UAV-to-UAV link's traffic load, a much larger ground-users can successfully send the data by using End-to-End link. In case of the relatively higher number of ground-nodes in a UAV Network, the proposed trajectory decision scheme changes the radius of circular trajectory unlike the even UAV trajectory deployment.

4.2 Convergence Speed and End-to-End Link Connection Probability

In this experiment, we measure the convergence speed of the proposed scheme. In the simulation, we set the nodes distribution according to the uniform, gaussian cluster distributions. The gaussian cluster distribution is the one that occurs in disaster areas where nodes gather in clusters. This behavior is in accordance with people gathering in refuge areas. We assume that the calculation of trajectory by control station takes 0.5 s. Figure 4 shows the convergence of proposed algorithm and the End-to-End link connection probability. In our proposed scheme, from the initial UAVs' trajectories which is evenly distributed, UAVs change the trajectory, which is assigned by the control station, with value of flying speed, v. Therefore, the UAVs' flying speed is one factor that influences convergence speed. Moreover, according to Fig. 4, we can get to know maximum End-to-End connection probability is changed with the nodes' distribution.

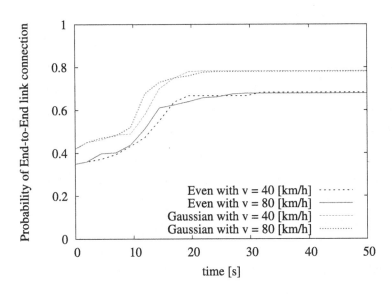

Fig. 4. Convergence speed of proposed UAV's trajectory decision scheme.

4.3 End-to-End Link Disconnection Duration

In this experiment, we measure the End-to-End link disconnection duration. To get know the character of link disconnection, we adopt an even UAV deployment scheme. Figure 5 shows the average End-to-End link disconnection duration. As shown in Fig. 5, the duration from link disconnection to link connection is related to UAV's speed and the number of UAVs. Moreover, it is considered that the radius also affects the End-to-End link disconnection duration. We need to take into consideration about the disconnection duration depending on the applications. This metric is most important when small delay is required application such as VoIP.

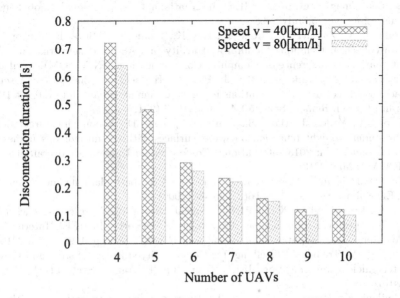

Fig. 5. Effect on the duration of disconnect.

5 Conclusion

In this paper, we proposed the UAV's trajectory decision scheme to improve the probability of End-to-End connection. At first, we evaluate the UAV-to-UAV link performance affected by UAVs' circular trajectory. Proposed UAVs' trajectory decision scheme change the center position vectors of circular trajectory and the radius of circular trajectory by using evaluated metric. Additionally, the proposed scheme decide to not make user outside UAV network. From the results, we confirmed that the proposed scheme achieves the low End-to-End delay trajectory.

Acknowledgement. This work was conducted under the national project, Research and Development on Cooperative Technologies and Frequency Sharing Between

Unmanned Aircraft Systems (UAS) Based Wireless Relay Systems and Terrestrial Networks, supported by the Ministry of Internal Affairs and Communications (MIC), Japan.

References

1. Johnson, A., Montgomery, J., Matthies, L.: Vision guided landing of an autonomous helicopter in hazardous terrain. In: 2005 IEEE International Conference on Robotics and Automation (ICRA), pp. 3966–3971, April 2005
2. Mandl, D., Sohlberg, R., Justice, C., Ungar, S., Ames, T., Frye, S., Chien, S., Tran, D., Cappelaere, P., Sullivan, D., Ambrosia, V.: A space-based sensor web for disaster management. In: 2008 IEEE International Geoscience and Remote Sensing Symposium, vol. 5, July 2008
3. Abdulla, A., Zubair, F., Hiroki, N., Nei, K., Fumie, O., Miura, R.: An optimal data collection technique for improved utility in UAS-aided networks. In: IEEE International Conference on Computer Communications (INFOCOM), April 2014
4. Crocker, R., Maslanik, J., Adler, J., Palo, S., Herzfeld, U., Emery, W.: A sensor package for ice surface observations using small unmanned aircraft systems. IEEE Trans. Geosci. Remote Sens. **50**(4), 1033–1047 (2012)
5. Kwon, H., Yoder, J., Baek, S., Gruber, S., Pack, D.: Maximizing target detection under sunlight reflection on water surfaces with an autonomous unmanned aerial vehicle. In: 2013 International Conference on Unmanned Aircraft Systems (ICUAS), May 2013
6. Tisdale, J., Kim, Z., Hedrick, J.: Autonomous UAV path planning and estimation. IEEE Robot. Autom. Mag. **16**(2), 35–42 (2009)
7. Lu, R., Lin, X., Shen, X.: SPRING: a social-based privacy-preserving packet forwarding protocol for vehicular delay tolerant networks. In: IEEE International Conference on Computer Communications (INFOCOM), pp. 1–9, March 2010
8. Kesting, A., Treiber, M., Helbing, D.: Connectivity statistics of store-and-forward intervehicle communication. IEEE Trans. Intell. Transp. Syst. **11**(1), 172–181 (2010)
9. Almi'ani, K., Viglas, A., Libman, L.: Energy-efficient data gathering with tour length-constrained mobile elements in wireless sensor networks. In: 2010 IEEE Conference on Local Computer Networks (LCN) (2010)
10. Keung, G., Li, B., Zhang, Q.: Message delivery capacity in delay-constrained mobile sensor networks: bounds and realization. IEEE Trans. Wireless Commun. **10**(5), 1552–1559 (2011)
11. Shah, R., Roy, S., Jain, S., Brunette, W.: Data MULEs: modeling a three-tier architecture for sparse sensor networks. In: 2003 IEEE International Workshop on Sensor Network Protocols and Applications, May 2003
12. Zhao, W., Ammar, M.: Message ferrying: proactive routing in highly-partitioned wireless ad hoc networks. In: 2003 IEEE Workshop on Future Trends of Distributed Computing Systems, May 2003
13. Wu, J., Yang, S., Dai, F.: Logarithmic store-carry-forward routing in mobile ad hoc networks. IEEE Trans. Parallel Distrib. Syst. **18**(6), 735–748 (2007)
14. Goddemeier, N., Daniel, K., Wietfeld, C.: Role-based connectivity management with realistic air-to-ground channels for cooperative UAVs. IEEE J. Sel. Areas Commun. **30**(5), 951–963 (2012)

15. Beard, R., McLain, T.: Multiple UAV cooperative search under collision avoidance and limited range communication constraints. In: 2003 IEEE Conference on Decision and Control, vol. 1, December 2003
16. Li, M., Wan, P.-J., Frieder, O.: Coverage in wireless ad hoc sensor networks. IEEE Trans. Comput. **52**(6), 753–763 (2003)

Hybrid Satellite-Aerial-Terrestrial Networks for Public Safety

Ying Wang$^{(\boxtimes)}$, Chong Yin, and Ruijin Sun

State Key Laboratory of Networking and Switching Technology,
Beijing University of Posts and Telecommunications, Beijing 100876,
People's Republic of China
wangying@bupt.edu.cn

Abstract. Wireless communication technologies play an irreplaceable role to satisfy Public Protection and Disaster Relief (PPDR) operational needs in the emergency situations. The existing practical solutions for PPDR system mainly include the dedicated public safety network, the commercial LTE network and the mobile satellite system (MSS), which are all separately operated due to the lack of a unified arrangement. In this context, this paper proposes a novel solution framework for the large-scale emergency scenarios, which is called the Hybrid Satellite-Aerial-Terrestrial (HSAT) system. The proposed HSAT system considers the integration of terrestrial components and the satellite network, and also added the low altitude platform (LAP) as a complementary component. Moreover, some new technologies in LTE are also included in the system, aiming to support the increasingly data-intensive traffic. By combining the respective advantages of each network, the proposed HSAT system can potentially offer higher throughput, wider coverage and stronger robustness, which are all highly demanded in PPDR networks.

Keywords: Public safety · Mobile satellite system · Low altitude platform · LTE network

1 Introduction

An effective Public Protection and Disaster Relief (PPDR) system is crucial to a successful response to emergency and disaster situations. Unlike the traditional communications in cellular networks, the PPDR system has a mission-critical aspect and thus places some special requirements on the underlying radio technologies. For example, the PPDR system should be easy to deploy, highly reliable, relatively low in price and high capacity-coverage.

Currently, two terrestrial wireless communication networks are utilized for emergency communication, i.e. the commercial cellular network and the dedicated public safety network (e.g. TETRA, APCO25 or DMR) [1,2]. The dedicated network aims at providing immediate access to the network with guaranteed reliability while the cellular network (e.g. LTE) is used to provide

© ICST Institute for Computer Sciences, Social Informatics and Telecommunications Engineering 2016
I. Bisio (Ed.): PSATS 2014, LNICST 148, pp. 106–113, 2016.
DOI: 10.1007/978-3-319-47081-8_10

some broadband data-centric services. In addition, as a complement to terrestrial mobile communication systems, the mobile satellite communication system (MSS) has proved to be a valuable gap filler in public safety networks, since it can provide services in the regions where terrestrial network collapses due to the disaster. Although MSS provides wider coverage and is more disaster tolerant, it usually requires the existence of Line of Sight (LOS) and endures longer transmission delay. Therefore, the integration of MSS and the terrestrial network becomes highly demanded. Moreover, during a large-scale natural disaster, the LTE base stations (BSs) could become overloaded or even totally destroyed. In these scenarios, the airborne communication systems have been recently studied for providing rapidly deployable and resilient accesses [3,4]. The aerial station is an air balloon or aircraft based low attitude platform (LAP), which can be built within one hour. It can not only play a role as the LTE base station, but also communicate with the satellite.

In this paper, we propose a solution framework for a large-scale PPDR network, which is called the Hybrid Satellite-Aerial-Terrestrial (HSAT) system. This system intelligently combines the satellite communication, the terrestrial network and the proposed LAP concept. Besides, some new technologies, such as the Device-to-Device (D2D) communication and cognitive radios (CR) will be added to the PPDR networks, in order to support the increasingly data-intensive traffic. Compared with the existing PPDR network, the proposed hybrid Satellite-Aerial-Terrestrial system could effectively combine the respective advantages of each network, and will provide a complete and feasible solution for the large-scale natural disasters.

The rest of the paper is organized as follows. In Sect. 2, we describe the existing terrestrial communication system, which mainly includes the dedicated public safety network and the LTE network. We also discuss the architecture of MSS for PPDR service provisioning. Then in Sect. 3, we will illustrate the basic architecture of the proposed HSAT system and discuss some critical challenges of the practical operation of the system. Finally, Sect. 4 concludes the paper.

2   Existing Public Safety Networks

2.1   Terrestrial Communication Networks

During a large-scale emergency or disaster, the public communication network may cease to work after suffering great damages. The dedicated public safety network for PPDR communication plays an essential role in the field first aid. Compared with the common mobile communication system, the dedicated network is devoted for special command and schedule. With the large-cell and low-density network organization, a base station can cover the range of tens of kilometers. If communication is interrupted in one region, a mobile station such as a vehicle station can continue to function for a certain area. The major feature of the dedicated network is that it can provide a rich set of voice-centric services, such as push-to-talk, group calling and emergency dispatching.

Nevertheless, with the prevalent of smart devices, the emergency information tends to be diversified, transferring from voice-centric to data-centric, the rescuing image or video information related geographic location is an example. Although some efforts have been made to improve the system capacity, the solution lags far behind the achievements of existing mobile networks. In addition, the commercial LTE system serves for ordinary users rather than only first responders. Thus, the adoption of commercial LTE technology for the PPDR community could meet common subscribers routine communications who are in the hard-hit area.

Although these two networks are operating separately from now on, some researches have been done to introduce the commercial LTE system to PPDR. The synergies between these two networks are obvious, which includes maximization of the economies of scale, better capacity, enhanced resiliency and improved radio coverage. Based on the above considerations, the integration of dedicated public safety network and LTE systems is irresistible future tendency.

2.2 Mobile Satellite Communication Networks

Based on the space platform (e.g. geostationary satellite, middle/low orbit satellite), satellite communication system is used for real-time acquisition, transmission and process of the spatial information. Due to the advantages of high disaster tolerance, large coverage and flexible network organization, the satellite communication system can communicate directly with the disaster acquisition system, the rescue command system and the disaster broadcasting system through vehicle or aircraft stations.

Satellite communication services can be categorized as fixed satellite services (e.g. satellite TV services) and mobile satellite services. The mobile satellite services are widely used in the marine, aviation, remote areas as well as disaster or emergency situations. During a disaster, satellite-based phone can communicate directly with the satellite system when the LOS is satisfied. Otherwise, the emergency communication vehicles or aircrafts would serve as a simple repeater that fills the NLOS holes. In this case, the vehicle or aircraft stations can retransmit the received signal at the frequency same as the satellite or not. As shown in Fig. 1, the satellite-based public safety network infrastructure consists of not only the satellite but also the vehicle/aircraft station. In fact, the introducing of the vehicle/aircraft station could bring many benefits, including filling the gaps in satellite coverage, enhancing the satellite capacity and improving the successful access probability.

So far, the network protocol of satellite system is developing from PPP, ATM to IP technology over satellite, aiming to connect with the terrestrial interoperability. Meanwhile, satellite service is also converting from a single service to integration with telecommunication and the internet. Thus, to meet the users massive and diversified multimedia requirements, the broadband and IP-based architecture becomes an irresistible trend.

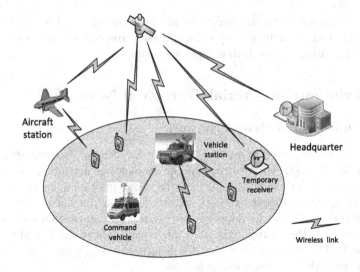

Fig. 1. The satellite-based public safety network infrastructure

2.3   Comparison of MSS and Terrestrial Networks

The above two subsections presented the basic characteristics of the terrestrial and satellite network respectively. In this subsection, we focus on the comparison of the two networks, so as to provide a clear direction for the design of system architecture. A brief outline of the comparison is shown in Table 1.

Table 1. Comparison of satellite and terrestrial systems

	Advantages	Disadvantages
Satellite	large coverage	require LOS transmission
	less impacted by disaster	long delay
		limited onboard power
Terrestrial	enhanced capacity	susceptible to disaster
	diversified multimedia services	

MSS could provide wider coverage and is more disaster tolerant. However, some limitations of satellite communications should not be ignored. For example, the satellite communication requires the existence of LOS so that users in shadowing or indoor areas cannot be covered effectively. Moreover, the satellite communication endures long transmission delay [5], which makes the dynamic resource allocation and Adaptive Modulation and Coding (AMC) no longer effective in satellite communication system. Other limitations may include the high cost and the incompatibility with LTE networks. On the other hand, the

LTE-based terrestrial system could provide low cost coverage for high-density populations. It also has higher spectrum efficiency and could satisfy a wide range of data communication needs in emergency scenarios.

3   Hybrid Satellite-Aerial-Terrestrial Network

3.1   Architecture of Hybrid Network

As described above, satellite system is characterized by large coverage and less impacted by disaster, while terrestrial communication network is featured as enhanced capacity and supportability of diversified multimedia services. Therefore, the integration of these two networks can bring significant benefits. Within the terrestrial networks, the adoption of commercial mainstream LTE technology to deal with the increasingly data-intensive applications is widely agreed within the PPDR community. Some promising technologies that could be useful in PPDR networks are described as follows.

1. All-IP system architecture and flexible air interface
    The LTE IP connectivity services are implemented by the Evolved Packet Core (EPC), which is the fundamental part of an LTE network. Within the EPC, the LTE could also offer different levels of interoperability provide a wide range of multimedia services and guarantee prioritized handling of emergency calls. The air interface supports flexible carrier bandwidths from below 5 MHz up to 20 MHz, which can satisfy different end-to-end QoS for different users.
2. D2D (Device-to-Device) Technology
    D2D communication allows adjacent devices within or outside of cellular coverage to communicate directly instead of relaying by the BS. It can largely offload the burden of BS, especially when only few BSs survived in the emergency situation [1]. Besides, the D2D communication has the potential to save the transmission power of terminals, which significantly enlarges the network lifetime in emergency scenarios.
3. CRs (Cognitive Radios)
    CR technology is expected to detect the spectrum hole and use the unoccupied spectrum opportunistically, thus improving the spectrum efficiency. It can also sense users needs through learning algorithm and allocate just enough radio resources for them [6]. Another aspect of CR is to develop applications that can locate, communicate and reach the victims who are stuck in disaster areas or behind obstacles.

Although the LTE network has great potential in broadband data service provisions, it is highly possible that most of the LTE BSs become damaged or even cease to work due to the severe natural disaster. In this case, the utilization of LAP as an alternative to the terrestrial BS will become necessary. Based on the above considerations, the complete architecture of our proposed HSAT system is shown in Fig. 2. As we can see, the HSAT system comprises three main parts:

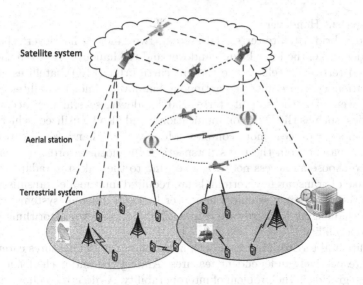

Fig. 2. The HSAT network architecture

the MSS, air balloon or aircraft based aerial station and the survived terrestrial network. The aerial station can be fast built and fill the gaps of destroyed LTE base station. It is generally about 100–1000 m high, lifting with a low-complexity LTE-A eNB, named Aerial eNodeB (AeNB). It is worth mentioning that the AeNB is expected to communicate with satellite as well as terrestrial users that within its coverage area.

In case of a large-scale natural disaster, some mobile BSs along with the public safety personnel will be sent to the incident areas. Such dedicated public safety network is specifically useful in providing robust and low-delayed services for the first responders. Meanwhile, the stationary LTE BSs are used to deal with the routine data traffic. Some new technologies, such as the Device-to-Device (D2D) communication and cognitive radios (CR) could also be added to the LTE network, in order to support the increasingly data-intensive traffic. In addition, we apply the LAP in the proposed architecture as an effective alternative to the terrestrial BSs, in case that the terrestrial BSs collapse and cease to work due to the disaster. Finally, as for the areas that are out of terrestrial coverage, the MSS could be utilized to provide seamless services for both the victims and first responders.

3.2   Integration Challenges

Although it is of significant benefits to simultaneously utilize the terrestrial and satellite network in the PPDR system, how to integrate the two networks effectively still faces some practical challenges. From the technical perspective, the major challenges may include the transparent handover issues, interpretability and resource allocation [7].

1. Transparent Handover

   In the hybrid satellite-terrestrial system, two kinds of handovers should be considered, i.e. the inter-beam handover and the handover between the satellite and terrestrial cells. Since the low earth orbit (LEO) satellites are nongeostationary, there may be continuous handovers among satellites even for fixed users. For the satellite-terrestrial handover, several new access technologies, such as the D2D communication could also be utilized, which makes the handover process more complex. Moreover, the handover is expected to be transparent from the users perspective. In another words, user ends are able to choose an access network according to the link state adaptively. For the sake of transparency, terminals are required multimode transmission and supports low latency seamless handover between different systems, which is much challenging for hardware chip design and handover algorithms [8].

2. Interoperability

   Satellite and territorial system both have their own architectures catering for their respectively independent features. Among the major challenges facing the integration is the problem of interoperability. As described above, in order to fulfill diversified public communication services, both of the two networks are developing towards broadband and IP. Meanwhile, IP protocol is so successfully used in the internet. Thus, constructing the integrated PPDR over IP is the future trend.

3. Resource Allocation

   There are some differences between the satellite and terrestrial communication systems, such as the standards, the forms of physical resource, frequency bands as well as the transmit power, coverage, capacity and control plane overhead, etc. Therefore, it is critical for the integrated system to utilize the limited radio resources effectively and efficiently. The resource allocation problem mainly consists of spectrum management and power control. Spectrum management: The satellite systems generally use s-band (2–4 GHZ) for direct communication with users, and Ka-band (27–40 GHZ) for feedback to the satellite receiver. The terrestrial system uses 900 MHz and 1800 MHz. Possibly, there still exists frequency overlapping. A series of measures should be used for efficient spectrum management, such as rational planning [8] interference avoiding or suppression [9] cognitive radio technology [10], etc. Power control: Due to the limited carrying capacity, the LAP is usually unable to carry enough battery, which is also true with the space based satellites. However, it is convenient for the ground based LTE BS to supply power. Therefore, different transmit power characters should be taken into consideration for extending the network effective lifetime.

4 Conclusion

The LTE-based terrestrial communication network has enhanced capacity and is expected to offer diversified multimedia services in an emergency situation. The satellite communication network could provide wider coverage and suffer less from the disaster. Therefore, the integration of these two networks could bring synergic gain. In this paper, we propose the HSAT system for PPDR networks, which combines the satellite communication, the terrestrial network and the proposed LAP concept. Besides, some new technologies, such as D2D and CR are added to the hybrid system. The proposed HSAT system can potentially offer higher throughput, wider coverage and stronger robustness, which is promising to be used in the PPDR networks.

Acknowledgment. This work is supported by National 863 Project (2012AA01A 50604) and National Nature Science Foundation of China (61121001).

References

1. Doumi, T., Dolan, M.F., Tatesh, S., Casati, A., Tsirtsis, G., Anchan, K., Flore, D.: LTE for public safety networks. IEEE Commun. Mag. **51**(2), 106–112 (2013)
2. Ferrs, R., Sallent, O., Baldini, G., Goratti, L.: LTE: the technology driver for future public safety communications. IEEE Commun. Mag. **51**(10), 154–161 (2013)
3. Bucaille, I., Hethuin, S., Munari, A., Hermenier, R., Rasheed, T., Allsopp, S.: Rapidly deployable network for tactical applications: aerial base station with opportunistic links for unattended and temporary events absolute example. In: Military Communications Conference, MILCOM 2013, pp. 1116–1120. IEEE Press (2013)
4. Gomez, K., Rasheed, T., Reynaud, L., Bucaille, I.: Realistic deployments of LTE-based hybrid aerial-terrestrial networks for public safety. In: IEEE 18th International Workshop on Computer Aided Modeling and Design of Communication Links and Networks (CAMAD), pp. 233–237 (2013)
5. Siyang, L., Fei, Q., Zhen, G., Yuan, Z., Yizhou, H.: LTE-satellite: chinese proposal for satellite component of IMT-advanced system. China Commun. **10**(10), 47–64 (2013)
6. Al-Hourani, A., Kandeepan, S.: Temporary cognitive femtocell network for public safety LTE. In: 2013 IEEE 18th International Workshop on Computer Aided Modeling and Design of Communication Links and Networks (CAMAD), pp. 190–195 (2013)
7. Yang, L., Zhang, Y., Li, X., Gao, X.: Radio resource allocation for wideband GEO satellite mobile communication system. In: International Conference on Wireless Communications and Signal Processing (WCSP), pp. 1–5 (2013)
8. Sadek, M., Aissa, S.: Personal satellite communication: technologies and challenges. IEEE Wirel. Commun. **19**(6), 28–35 (2012)
9. Deslandes, V., Tronc, J., Beylot, A.-L.: Analysis of interference issues in integrated satellite and terrestrial mobile systems. In: 2010 5th Advanced Satellite Multimedia Systems Conference (ASMA) and the 11th Signal Processing for Space Communications Workshop (SPSC), pp. 256–261. IEEE Press (2010)
10. Sharma, S.K., Chatzinotas, S., Ottersten, B.: Cognitive radio techniques for satellite communication systems. In: Vehicular Technology Conference (VTC Fall), pp. 1–5. IEEE Press (2013)

Satellites, UAVs, Vehicles and Sensors for an Integrated Delay Tolerant Ad Hoc Network

Manlio Bacco[1,3]([envelope]), Luca Caviglione[2], and Alberto Gotta[1]

[1] Information Science and Technologies Institute (ISTI),
National Research Council of Italy (CNR), Pisa, Italy
felice.bacco@unisi.it, alberto.gotta@isti.cnr.it
[2] Institute of Intelligent Systems for Automation (ISSIA),
National Research Council of Italy (CNR), Genova, Italy
luca.caviglione@ge.issia.cnr.it
[3] Department of Information Engineering and Mathematic Science,
University of Siena, Siena, Italy

Abstract. In this paper, a fully meshed mobile ad hoc network is introduced as an alternative to a classical wide area network, as the Internet. Internet of Things, Internet of Vehicles, Unmanned Aerial Vehicles and satellites are the enabling technologies of such a complex scenario with support to multilevel mobility, overlaid deployments, as well as techniques offering Delay Tolerant Networks services.

In this perspective, the paper provides an insight of the most relevant technological issues to guarantee a proper Quality of Service level between an Unmanned Aerial Vehicle communicating with a remote data center through a satellite link. Besides, this work also evaluates the coverage of such a concept in a metropolitan area.

Keywords: Satellite · UAV · IoT · IoV · DTN

1 Introduction

By offering a vast set of services and informative contents, Internet is the worldwide public computer network representing the main medium for mass communication. In the last few years, desktops have been progressively replaced by mobile devices, like smartphones and tables. This trend culminates in many small embedded devices rapidly increasing their presence in the Internet. To this aim, the main technological enabler is the recent evolution of protocol stacks like the IPv6 over Low power Wireless Personal Area Networks (6LoWPAN) [1], which enables to expose a single-tiny device as a node of the Internet. Potentially, 7 billion of users will lead to 10 billions of connected devices: such a vision is commonly called the Internet of Things (IoT).

Another important component is given by ad-hoc networks allowing a device using a wireless link to reduce its utilization of costly data plans to access the

© ICST Institute for Computer Sciences, Social Informatics and Telecommunications Engineering 2016
I. Bisio (Ed.): PSATS 2014, LNICST 148, pp. 114–122, 2016.
DOI: 10.1007/978-3-319-47081-8_11

Internet. The latter are usually too expensive (or not required) compared to the amount of data to be delivered.

While the state-of-the-art literature on ad-hoc networks spans over almost two decades, still, no actual applications use them, primarily due to the poor support of real-time requirements. However, owing to the ferment around the IoT paradigm, jointly with the proliferation of smart and mobile devices, the market for ad-hoc products is becoming relevant. Another important input for the adoption of ad-hoc mechanism is given by the lack of cost-reduction of using wireless accesses offered by Internet Service Providers (ISPs) or Telcos. Thus, implementations of frameworks based on ad-hoc principles are becoming commercially available (see, e.g., [3,4]). Especially, the actual IoT panorama is populated by several monitoring applications to create large data-sets (commonly defined as Big Data) used to elaborate new business or relation models.

Therefore, developing an alternative "internet" to assure a proper internetwork is a convenient solution to avoid grounding on the Internet. We like to name such a deployment *Alternet*. However, connectivity is not the only requirement to be satisfied. In fact, when in presence of high delays and unattended ad-hoc installations, a proper protocol architecture must be available, also to achieve large-scale coverages. Thus, the Delay Tolerant Network (DTN) framework is the preferred solutions to satisfy these constraints.

In this perspective, this paper deals with the characterization of a DTN-based application used in the context of environmental monitoring for smart cities [5]. The key technological challenge consists in building a mobile ad-hoc environmental monitoring network, interconnected by opportunistic links created between public urban vehicles and Unmanned Aerial Vehicles (UAV), i.e., drones. Moreover, a satellite link enables the drone to exchange data towards the remote data center. Performance metrics have been obtained through on-field experiments with a real UAV built by the Information Science and Technologies Institute (ISTI) of the National Research Council of Italy (CNR) [6]. Nevertheless, to understand whether an UAV can be actually deployed in real urban scenarios, simulations characterizing an urban area (i.e., Pisa) are also provided.

The main contribution of this paper is the evaluation of mechanisms to provide a proper degree of reliability when exchanging data, despite the delays or phenomena of intermittent connectivity.

The remainder of the paper is structured as follows: Sect. 2 provides an overview of the reference scenario, while Sect. 3 discusses the setup of simulations. Section 4 showcases numerical results, and Sect. 5 concludes the paper.

2   Reference Scenario

The scenario considered in this work for Alternet consists in a set of "islands" of mobile and fixed ground nodes, such as public vehicles and hotspots. Nodes are not always connected, since are spread in the urban area. Yet, they can communicate when in proximity, as a consequence of their mobility. This usually happens at periodic inter-meeting times between mobile-mobile and mobile-fixed nodes.

The positions of mobile nodes are not known a-priori. Moreover, their routes could never collide, leading to nodes (vehicles) isolated from the rest of the network. As a consequence, some entities will never be able to send data to the remote sink (i.e., the data center). To solve this issue, we use an UAV, which can connect both mobile and fixed nodes with the remote sink.

Due to the scenario, the flying path of the UAV must be large enough to offer the maximum probability of encountering as many nodes as possibles, despite their positions. Then, the UAV periodically returns to the headquarter, where a satellite access point provides a reliable backbone to the dedicated data collection service. Obviously, this methodology can be straightforwardly extended for the case of multiple UAVs.

As regards the method to provide connectivity (with the acceptation of data routability) among fixed, ground mobile and aerial vehicles, a pure epidemic routing protocol [8] is assumed available. Besides, each node composing the architecture is equipped with short-range and high-bandwidth air interfaces. From the physical layer viewpoint, the network is assumed to be mostly disconnected, thus only intermittent short-range links are available.

Figure 1 depicts the map of the city of Pisa, as well as the geographical locations of fixed nodes, the sink and the paths of mobile nodes. Two different paths for the UAVs flights are considered, as to evaluate different metrics, such as the packet delivery ratio, the delays, and the hop count. In our scenario, a very low data rate is supposed to be produced by a sensor source. In fact, according to the aim of the project [5], a telemetry of pollutant and climatic factors is collected firstly to provide a daily and georeferred bulletin. Secondly, another objective is to populate a historical data base to correlate these values with the incidence of cardiovascular pathologies on citizens.

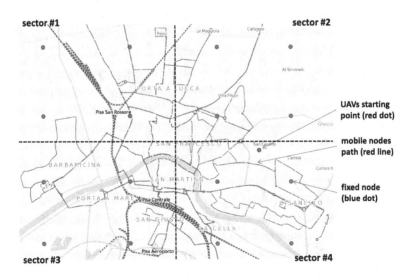

Fig. 1. The map of Pisa used as a reference scenario.

According to the nature of the monitored data, each fixed or mobile sensor node (except the UAVs) generates a 100 byte Protocol Data Unit (PDU) every 5 min. The packets lifetime is equal to one hour and maximum two retransmission are allowed.

Lastly, Fig. 2 portraits the UAV used in our preliminary round of tests used to properly set up the simulator.

Fig. 2. The UAV used in our preliminary round of tests.

3   Simulation Setup

In order to evaluate the Alternet-based architecture proposed in this work, a thorough simulation campaign has been performed. To this aim, a map of $4000 \times 4000\,\mathrm{m}^2$ is considered (built on the topography of Pisa, as depicted in Fig. 1), where fixed nodes are deployed with a regular placement, and mobile nodes can move along predefined routes. The physical space is ideally divided in four equal sectors and four UAVs are used, each of them covering a sector, therefore they can never encounter another UAV while in flight. The UAVs have their recharge base station (the red dot in Fig. 1) where a satellite relaunch is present and devoted to deliver the buffered data from the UAVs to the remote data center. The flight duration is around 15 min, which corresponds to the mean battery lifetime; thus, every 15 min the UAVs leave the base and start their flight in the designed sector after a battery replacement.

In such a scenario, we consider *four* different possibilities to exchange PDUs:

1. a mobile node encounters another mobile node;
2. a mobile node encounters a fixed node;
3. a UAV encounters a mobile node;
4. a UAV encounters a fixed node.

Since an epidemic routing protocol is assumed, when a node encounters another node in its path, it transmits all the packets in the buffer, even if the other node is not the final destination. In this way, the data propagate through the network trying to reach the final target through multiple and different paths. To avoid the saturation both of network resources and buffers, if the data successfully reaches the sink (i.e., the final destination), an *antipacket* is transmitted back to the source node.

In essence, the antipacket is a sort of "receipt" to the source node (see, e.g., the VACCINE mechanism [9]), which triggers the deletion of the acknowledged data from the buffers. Hence, each PDU has a proper ID. Additionally, the antipacket gives the "immunity" to nodes, as to prevent the uncontrolled propagation of unneeded data. An antipacket lifetime is equal to the residual lifetime of the packet with the same ID.

More copies of a PDU with a given ID can reach the sink. In this sense, a routing protocol based on epidemic data dissemination might not be the best choice. However, at this stage, its adoption offers three main advantages: (*i*) its implementation is simple and does not require additional overheads for path discovery or for exchanging Global Positioning System (GPS) coordinates (as it happens in geographic routing); (*ii*) since our scenario implements a totally meshed network exchanging tiny PDUs with a low generation rate, even in presence of duplicates, saturation is an unlikely event; (*iii*) data loss is very low because of the high redundancy.

Regarding the parameters characterising our simulated environment, they have been collected in a preliminary set of trials performed with an UAV in the Pisa Research Area of the CNR. Specifically, equipping nodes with an IEEE 802.11g air interface leads to a range of ~130 m, as presented in [10] and as confirmed by the measurement campaigns performed with our UAV in the CNR research campus.

The maximum speed for UAVs is 8 m/s. They are equipped with a GPS navigator and can be remotely controlled in a range of 2 km or programmed to follow a GPS route. Since the second modality is more appealing because does not require any human interaction, we did test by implementing two different autonomous flight strategies:

- **random walk**: the UAV randomly deviates from its route, allowing to have a more vast coverage of the sensing area during the flight period;
- **planned way-points**: the UAV follows the fixed routes of ground vehicles (e.g., streets). In this case, PDUs generated by mobile nodes typically have to cross an additional hop toward a fixed node before reaching the UAV.

The proposed scenario has been simulated by using ns-3 [7] and the DTN protocol implementation for ns-3 in [2].

4   Numerical Results

Six different scenarios have been used during the simulations, slightly different with regards to mobile nodes paths, to ensure enough randomness in the
movement, as well as the coverage of the whole city area.

Figure 3 shows data and control traffic of one of the simulations. The
antipackets are sent back from the remote data center at the arrival of the
data packets, and it is visible that they are periodically generated, with a period
equal to the flight time of the UAV. The epidemic routing protocol is responsible
for the large number of packets in the network, even with a low data rate.

Figure 4 shows the goodput for the six scenarios: the planned UAV flight
ensures an higher delivery ratio (goodput) than the random walk, close to 0.9,
even if the difference between those is very low, about 0.05.

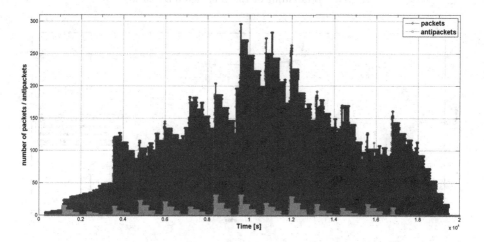

Fig. 3. Data and control traffic during a sample simulation

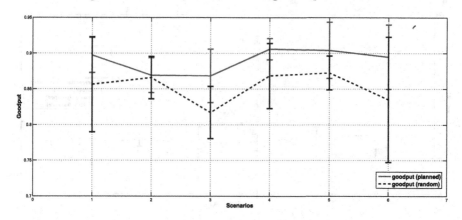

Fig. 4. Goodput of the considered scenarios

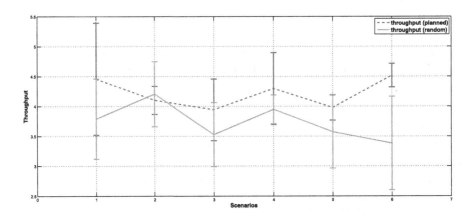

Fig. 5. Throughput of the considered scenarios

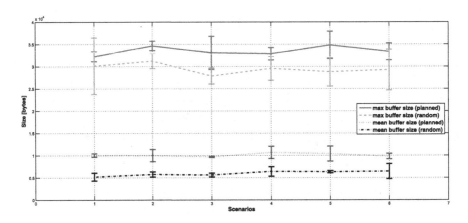

Fig. 6. Buffers Size of the considered scenarios

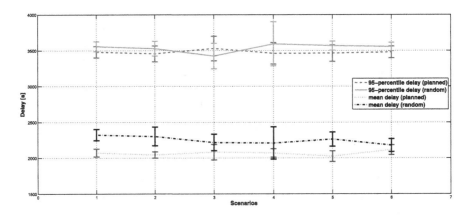

Fig. 7. Delivery Delay of the considered scenarios

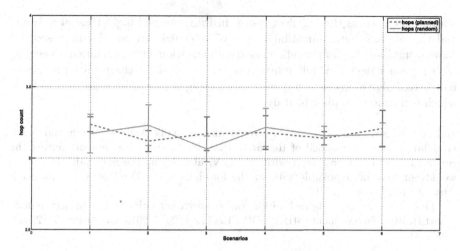

Fig. 8. Mean Hop Count of the considered scenarios

Each scenario shows an high number of duplicates reaching the sink. In fact, the throughput is from fourfold to fivefold higher of the goodput, as showed in Fig. 5. This can be explained by describing the mobile nodes behavior: a mobile node can encounter a certain number of fixed nodes and more than one UAV, because the path may be spread on more than one city sector. The epidemic routing protocol will continuously create packet duplicates, thus increasing the throughput to such an high value.

Figure 6 shows the buffer size of the nodes in the network: it is clearly visible that the planned UAV flight requires larger buffers with respect to the random flight. It is evident that, in the planned flight, an UAV will surely encounter, at least, the four fixed nodes in its sector plus a certain number of mobile nodes; in the random flight, the number of encountering events may be slightly inferior due to the randomness of paths, thus requiring less space in the buffers of the peers.

Figure 7 shows the mean delivery delay and the 95-percentile delivery delay of packets: the random flight has an higher delay, comparing to the planned flight. In the latter case, a greater number of packets is collected during each flight, as confirmed by the buffers size. Then, a higher number of packets is delivered to the sink at each flight with respect to the former case.

In Fig. 8 the mean hop count is shown: the hop count results almost the same for the planned and the random UAV flight, that is ≈ 3.2.

5   Conclusions

In this paper we presented a mobile ad-hoc network-based scenarios using UAVs to collect data produced by ground nodes, to be transferred via a remote sink through a satellite link. The results of the experiments are based on realistic parameters, such as the bus fleet routes in Pisa, the real flight time of a drone, and the wireless communication ranges of vehicular devices. As discussed, we showed the feasibility of providing the communication with a remote data center, with a proper degree of reliability. Also, the analysis of the results provides a metric to design a production-quality monitoring network in a smart city as Pisa, which is a candidate pilot in Italy.

Acknowledgements. The authors would like to thank Andrea Berton and Fabio Grassini of National Council of Research, Pisa, for the technical support during the (many) hours of test and flight with the UAV, also during the weekends. This work would not have been possible without the foresight of the Director of the Research Area of Pisa, Ing. O. Zirilli.

This work has been funded within the research activities of the project named Smart Healthy Environment (SHE), POR CReO Fesr, 2007–2013, CUP 6408. 30122011. 026000115.

References

1. 6LoWPANs, RFC4944 - Transmission of IPv6 Packets over IEEE 802.15.4 Networks, RFC6282 - Compression Format for IPv6 Datagrams over IEEE 802.15.4-Based Networks, and RFC6775 - Neighbour Discovery Optimization for IPv6 over Low-Power Wireless Personal Area Networks
2. Lakkakorpi, J., Ginzboorg, P.: ns-3 module for routing and congestion control studies in mobile opportunistic DTNs. In: Proceedings of the SPECTS 2013, Toronto, Canada, July 2013
3. open-garden. https://opengarden.com/
4. https://developer.apple.com/library/ios/documentation/MultipeerConnectivity/ Reference/MultipeerConnectivityFramework/Introduction/Introduction.html
5. Smart Monitoring Integrated System for a Healthy Urban ENVironment in Smart Cities (SmartHealthyENV). http://www.isti.cnr.it/research/unit.php?unit=WN& section=projects
6. National Research Council (CNR). http://www.area.pi.cnr.it/
7. ns-3. https://www.nsnam.org/
8. Vahdat, A., Becker, D.: Epidemic routing for partially connected ad hoc networks. Tech. Rep. CS-200006, Department of Computer Science, Duke University, Durham, NC (2000)
9. Haas, Z., Small, T.: A new networking model for biological applications of ad hoc sensor networks. IEEE/ACM Trans. Netw. **14**(1), 27–40 (2006)
10. Holland, G., Vaidya, N., Bahl, P.: A rate-adaptive MAC protocol for multi-hop wireless networks. In: Proceedings of the 7th Annual International Conference on Mobile Computing and Networking (ACM), pp. 236–251 (2001)

Smartphones *App*s Implementing a Heuristic Joint Coding for Video Transmissions Over Mobile Networks

Igor Bisio[✉], Fabio Lavagetto, Giulio Luzzati, and Mario Marchese

DITEN, University of Genoa, Genoa, Italy
{igor.bisio,fabio.lavagetto,mario.marchese}@unige.it
giulio.luzzati@edu.unige.it

Abstract. This paper presents the Heuristic Application Layer Joint Coding (Heuristic-ALJC) scheme for video transmissions aimed at adaptively and jointly varying both applied video compression and source encoding at the application layer used to protect video streams. Heuristic-ALJC includes also a simple acknowledgement based adaptation of the transmission rate and acts on the basis of feedback information about the overall network status estimated in terms of maximum allowable network throughput and link quality (*lossiness*). Heuristic-ALJC is implemented through two smartphone Apps (transmitter and receiver) and is suitable to be employed to transmit video streams over networks based on time varying and possibly lossy channels. A performance investigation, carried out through a real implementation of the *App*s over Android smartphones, compares Heuristic-ALJC with static schemes.

1 Introduction

The nature of the modern Internet is heterogeneous and implies the technical challenges of Quality of Service (QoS) guarantees and the quick deployment of new telecommunications solutions. These challenges need significant effort in the fields of the design of reliable and reconfigurable transmission systems, open source software, interoperability and scalability [1].

The mentioned internet scenario constitutes the reference for this paper: the considered network is characterized by radio and satellite links and includes mobile devices such as smartphones, employed to acquire and transmit video streams through dedicated *App*s. An applicative example of the considered environment concerns future safety support services: after a critical event (e.g., a road accident, a fire), first responders (e.g., a rescue team or just a person on site) can register a video by a smartphone and send it to an experienced operator over wireless/satellite heterogeneous network to allow managing rescue operations more consciously. In the described framework, static management of video compression and protection is not an optimal choice. Dynamic adaptation of video flow is necessary. It may be acted by opportunely tuning both the amount of data offered to the transmitting device and the amount of redundancy packets to

© ICST Institute for Computer Sciences, Social Informatics and Telecommunications Engineering 2016
I. Bisio (Ed.): PSATS 2014, LNICST 148, pp. 123–131, 2016.
DOI: 10.1007/978-3-319-47081-8_12

protect the video from losses. A possible improvement may derive by considering the impact that each of these tunings has on the other and by evaluating the joint effect of the two on the whole system performance. Following this scientific line, to guarantee a ready-to-use and satisfactory video fruition, two *Apps*, based on the Android OS, have been designed, implemented and tested. As described in the remainder of this paper, the *Apps*, a Transmitter *App* and a Receiver *App*, employ an application layer joint coding algorithm for video transmission based on a heuristic approach suited to be applied over smartphone platforms. The algorithm adaptively and jointly varies both video compression and channel coding to protect the video stream. It operates at the application layer and it is based on the overall network conditions estimated in terms of network maximum allowable throughput and quality (packet cancellations or *lossiness*): on the basis of information about packet loss, a given protection level is chosen; in practice, the amount of information and redundancy packets is chosen. Established the amount of available information packets and estimated the maximum allowable network throughput, video compression is consequently adapted to assure the best quality. The proposed solution also includes a simple acknowledgement-based adaptation of the transmission rate at the application layer aimed at not losing information in the application layer buffers. The proposed application layer joint coder considers the underlying functional layers as a black box. The *Apps* do not need any knowledge about implementation details and do not require any intervention regarding the underlying layers. The framework behind this work has been preliminarily presented in [2] and described in detail in [3].

2 State of the Art and Aim of the Paper

[4] demonstrates the existence of two sub-spaces called performance regions, and shows that the employment of application layer coding is significantly advantageous in one region, while it is detrimental in the other one. The first performance region contains the systems that experience light channel errors and low packet loss probability. The second region contains the systems characterized by relevant channel errors. Referring to [4], the mentioned coding approach may improve the performance only in the systems with low packet loss probability due to channel errors because error prone channels require so high levels of redundancy that they cause packet losses due to congestion. A solution to this limit is proposed in [5, Chap. 1]: increasing protection does not result in an increased offered load because the packet transmission rate is kept constant or, as done in this paper, adapted to the estimated maximum network throughput. Controlling the overall packet transmission rate, the network load is under control but, increasing protection implies reducing the amount of sent information per time unit (e.g. the size of sent video frames) and, consequently, the quality of sent information. In other words, the impact on the network load is controlled, information is more protected against channel errors, but the information distortion increases and impacts negatively on the QoE. For this motivation, if this type of solutions are applied, an end-to-end distortion minimization algorithm should be devised, to

get a joint source-channel coding approach. For instance, as done in this paper, a proper compression level may be selected consequently to the choice of the protection level. [6] investigates a joint coding solution at the application layer assuming the traffic generated by Gaussian sources. The contribution of this paper is inspired by the cited literature but, to the best of the authors' knowledge, there is no investigation about real implementations of joint source-channel coding at the application layer. This paper considers video streams acquired by a smartphone. The implemented Android *Apps* are aimed at jointly compressing and protecting the video dynamically so to guarantee a good QoE of the received video in case of error prone channels, limiting the offered load to the network. To reach the aim, differently from the aforementioned approaches, we employ a method to prevent exceeding the maximum allowable network throughput and to estimate the packet loss. The benefit of the designed coding has been highlighted through real video transmissions with smartphones over an emulated network, similarly as done in [7].

3   Implemented *Apps*

3.1   Preliminarily Definitions

The implemented applications put into operation video streaming between two distinct smartphones based on the Android OS. We describe the two *Apps* (Transmitter and Receiver), the software architecture, and the related structures in the following.

The chosen source encoder for video frames is MJPEG. From the practical viewpoint, an MJPEG video flow is a series of individual JPEG coded pictures representing the video frames. Concerning channel coding, LDPC [8] has been chosen for its computational feasibility. The resolution for video frames is QCIF (Quarter Common Intermediate Format, 176×144 pixels). The source coder is implemented by Android's API through a Java object to compresses a raw image through JPEG by quality index (decided by the heuristic algorithm proposed in this paper) as an input. The LDPC codec has been taken from an existing implementation [9] by adapting the source code as a library of the Android Native Development Kit (NDK).

The sequence of information processing actions of MJPEG video frames may be described as follows. A single video frame (i.e. a JPEG coded picture) is a **content** that is identified by a unique **content id**. The video stream is composed of a sequence of video frames. As shown in the right part of Fig. 1, which shows also Heuristic-ALJC actions described in Sect. 4, each video frame is divided into **video packets** (each **video packet** contains, at most, one video frame) also adding a proper header H, described in detail in Subsect. 3.2. Video packets are stored in a **processing buffer** of fixed length (35 packets in this paper). Once the number of **video packets** in the buffer reaches a certain threshold (called channel coding threshold - CCT, dinamically managed by the heuristic algorithm introduced in this paper), the video packets contained in the buffer enter the LDPC coder that generates a number of **redundancy packets** suitable to fill the rest of the buffer. In practice, the threshold CTT decides the amount of packets dedicated to transmit video information and, consequently, the amount of redundancy packets. Both video and redundancy packets have a length of 1024 [byte]. The sequence of video packets and the related **redundancy packets** compose a **codeword** (of 35 packets, as said), identified by a **sequence number**. The stream of packets composing codewords is stored into a **codeword buffer** from where the UDP transport protocol picks up and transmits the packets. A single packet is the transmission unit handled by UDP. A feedback channel allows the receiver to send *report packets* back to the transmitter. It is used to obtain information about the channel status.

3.2   Application Layer Packet Header

A small amount of control data (i.e. a header) in order to allow decoding operations and rebuilding individual *contents* from the transport layer data flow has been added to video packets. It is composed of 24 [byte], six Java integers, and contains the following fields: **FEC**, the number of redundancy packets; **Content ID**, a progressive number that identifies to which *content* (i.e., frame) the payload data belongs to; **Codeword Number**, a progressive number identifying the *codeword* which the *packet* belongs to; **Sequence Number**, a progressive number that individuates the *packet* position within the *codeword*; **Content Size**, which specifies the number of bytes composing the *content*; and **Offset**, measured in bytes, which indicates the distance from the beginning of the *content* (i.e., the JPEG image) where the *packet*'s payload must be written when the *content* is rebuilt.

3.3   Transmitter and Receiver *App*

The transmitter *App* has the tasks: to acquire frames from the smartphone camera; to compress them by using JPEG; to perform LDPC-encoding; to queue codewords in the codeword buffer employed to regulate the transmission rate; and to deliver them to the UDP transport protocol. The transmitter app is composed of: *streamer*, performing data processing and transmission, and *listener*, managing the feedback information received by report packets. The *listener* enables the adaptive capabilities of the transmission, and exploits the feedback information to compute source-channel coding parameters and to adapt the transmission rate to the maximum allowable throughput, as explained in Sect. 4.

The receiver *App* has a similar structure: a *listener* is bound to a particular UDP port and stores the received packets. LDPC decoder acts when either *(i)* the reception of a *codeword* is complete or *(ii)* a packet belonging to a more recent *codeword* (i.e., a codeword with a higher *Codeword Number*) unexpectedly arrives. Once the content of the LDPC protected stream has been recovered, JPEG frames are rebuilt and sequentially displayed on screen. Whenever a decoding session is completed, a *responder* fills the associated *report packet* and sends it to the transmitter.

4   Heuristic-ALJC

Heuristic ALJC method proposed in this paper is aimed at solving heuristically the problem formally defined in literature and represents the algorithmic core of the implemented Transmitter *App*. The constraint R_0 and the packet loss probability ϱ_k are usually unknown *a priori* and need to be determined. Our heuristic ALJC solution is based on three phases: *(i)* transmission rate adaptation through the employment of the *report packets* at the application layer; *(ii)* selection of the channel coding parameters; *(iii)* selection of the source coding parameters.

Each *report packet* carries information about the number of lost packets for each codeword and is sent each time a codeword is received. In this way, the transmitter is aware of how fast the mobile network can deliver the video, i.e., the transmitter derives an estimation of the maximum network throughput currently available, and of how vulnerable to losses is the sent video in the process of traversing the entire network. Concerning transmission rate adaptation, the regulation is acted on the basis of the *report packet* reception that enables the transmission of further codewords. Once the report packet for a given codeword is received, the corresponding codeword is acknowledged and removed from the codeword buffer. In case report packets are missing or delayed, the consequence is that the codeword buffer may saturate. In this case the transmission of codeword packets stops until a new report packet arrives so avoiding losing packets in the codeword buffer but affecting the average transmission rate. The rationale on the basis of this rate adaptation scheme is that, assuming the return channel reliable, the missing/delayed reception of report packets is interpreted as errored/narrowband forward channel. In the case the transmission rate adapter should not take any action, the transmission rate is limited to one codeword each 10 [ms]. Being a codeword composed of W [byte] (35 packets of $1024 + 24$ [byte], for payload and header respectively, the maximum possible transmission rate is limited by the ratio $\frac{W}{10}$ [byte/ms]. Concerning the selection of source and channel parameters, a single parameter is employed in the implemented *Apps* for both source and channel coding. In the following, the mentioned parameters will be denoted as $s_{1k} = Q$ and $c_{1k} = R_c$. R_c is the ratio between the overall number of video packets (i.e. the CTT threshold) and the fixed codeword length W ($R_c = $ CTT threshold$/W$). R_c is computed by Heuristic-ALJC and passed to the LDPC channel coder. Established the CTT threshold value and estimated through the arrival frequency of report packets the maximum network throughput currently available, Heuristic-ALJC choses the best value of the JPEG coder quality index Q and passes it to the JPEG coder. The main actions performed by ALJC are evidenced in the left part of Fig. 1.

5   Performance Investigation

5.1   Testbed

We have implemented a testbed to emulate the reference scenario described in the introduction. Two separate Android devices, implementing the transmitter

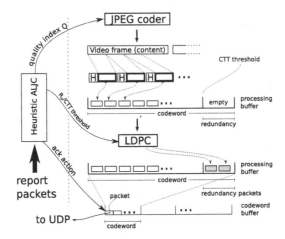

Fig. 1. Heuristic-ALJC actions

and receiver Apps, communicate through a WiFi local network connected to a machine that emulates the effect of a mobile network. On the receiving side, another WiFi network is used to interconnect the second device. The emulation machine is a regular PC running a Linux-based operating system, and implementing the *netem* tool to manage the outgoing traffic of each WIFI interface by tuning available channel bandwidth, packet loss, bit error rate (BER), and delay (fixed to 100 [mS] in all shown test).

5.2   Scenarios and Performance Metrics

Table 1 contains bandwidth and BER values for each emulated scenario.

In order to evaluate the performance, we have compared Heuristic-ALJC, implemented through the two designed Apps, with two opposite static policies assuring minimum protection/maximum quality ($R_c = 30/35$, $Q = 100$), and maximum protection/minimum quality ($R_c = 4/35$, $Q = 20$). The first group of tests evaluates Heuristic-ALJC behaviour during three minutes long sessions,

Table 1. Test scenarios

	Bandwidth	BER
A	400 Kbps	0 %
B	400 Kbps	10 %
C	400 Kbps	35 %
D	180 Kbps	0 %
E	180 Kbps	10 %
F	180 Kbps	35 %

for static channel conditions. A second group of tests investigates the system adaptation capabilities over time by varying network conditions. In order to measure the quality of individual frames of the MJPEG sequence, we utilize the Structural SIMilarity ($SSIM$) index, introduced in [10]. $SSIM(f_i, \widehat{f}_i)$ provides a quality measure of one of the frames ($\widehat{f}_i$) supposed the other frame (f_i) of perfect quality. $SSIM$ represents a good choice since it follows the Mean Opinion Score - MOS more closely than other indexes such as the Peak Signal to Noise Ratio (PSNR) and the Mean Square Error (MSE). $SSIM$ is computed over small portions of a frame, and the whole frame index $SSIM(f_i, \widehat{f}_i)$ is obtained by averaging the individual portion values. $SSIM$ index ranges from 0 (completely uncorrelated frames) to 1 (identical frames) and can be considered as a degradation factor. In order to evaluate the performance we have devised a performance index with the following requirements. It must reward high quality frames, a fluent video stream, and penalize corrupted or lost frames. Index I in (1) satisfies such requirements

$$I = \frac{\sum_{i=1}^{U} SSIM(f_i, \widehat{f}_i) \cdot f_{received}^{TOT}}{T_{sim}} \qquad (1)$$

and can be interpreted as a *quality-weighted average frame rate*.

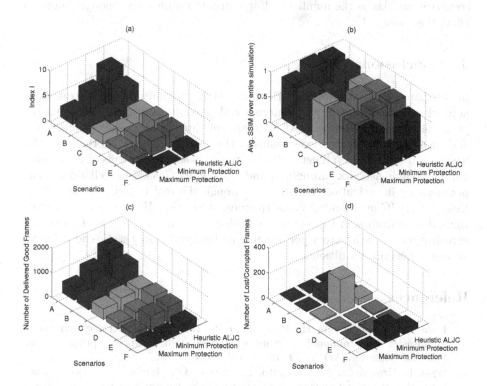

Fig. 2. Simulation of static channel behaviour

5.3 Performance Results

(1) Static Channel Scenarios: In this Section we show how our Heuristic-ALJC behaves when channel characteristics do not vary over time. Figure 2 shows the values of: Index I (a); average SSIM over the entire test (b); number of delivered good (decodable) frames (c); and number of lost/corrupted frames (d), for Heuristic-ALJC, Minimum and Maximum Protection schemes, for scenarios from A to F. The Maximum Protection scheme assures no loss (d) in all scenarios, even in 10 % BER (B and E) and 35 % BER (C and F) scenarios, but it dedicates so many packets to redundancy that the transmission rate of video frames is too reduced. This implies a limited number of delivered frames (c). Index I is low for any scenario. The Minimum Protection scheme behaviour may be satisfying for no loss scenarios, even if the large Q value imposed implies large frame size and consequent limited number of delivered frames (c), but it is highly inefficient for loss scenarios, where the large number of corrupted frames (d) heavily affects the quality (b) and consequently, Index I value (a). Heuristic-ALJC, by estimating the network available throughput over time, by tuning the protection level and adapting the source coding, always outperforms static solutions concerning Index I. It assures the highest number of successfully delivered frames (c) for all scenarios, and keeps the number of lost/corrupted frames low enough so not to affect the quality (b).

6 Conclusions

In this paper we have presented Heuristic-ALJC to transmit video streams on networks characterized by time varying and possibly lossy channels. From the practical viewpoint, Heuristic-ALJC adaptively applies both video compression and encoding to protect video streams at the application layer on the basis of a feedback about the overall network conditions, measured in terms of both maximum allowable network throughput and link quality (packet cancellations). The performance investigation, carried out through the real implementation of the Heuristic-ALJC over Android smartphones, shows that Heuristic-ALJC adapts the video transmission to network conditions so allowing an efficient resource exploitation and satisfactory performance and outperforming static coding under all tested network conditions.

References

1. Fouda, M.M., Nishiyama, H., Miura, R., Kato, N.: On efficient traffic distribution for disaster area communication using wireless mesh networks. Springer Wirel. Personal Commun. (WPC) **74**, 1311–1327 (2014)
2. Bisio, I., Grattarola, A., Lavagetto, F., Luzzati, G., Marchese, M.: Performance evaluation of application layer joint coding for video transmission with smartphones over terrestrial/satellite emergency networks. In: 2014 International Conference on Communications (2014, to appear)

3. Bisio, I., Lavagetto, F., Luzzati, G., Marchese, M.: Smartphones apps implementing a heuristic joint coding for video transmissions over mobile networks. Mob. Netw. Appl. **19**, 552–562 (2014)
4. Choi, Y., Momcilovic, P.: On effectiveness of application-layer coding. IEEE Trans. Inf. Theory **57**(10), 6673–6691 (2011)
5. Bovik, A.C.: Handbook of Image and Video Processing (Communications, Networking and Multimedia). Academic Press Inc., Orlando (2005)
6. Bursalioglu, O., Fresia, M., Caire, G., Poor, H.: Joint source-channel coding at the application layer. In: Data Compression Conference, 2009, DCC 2009, pp. 93–102, March 2009
7. Martini, M., Mazzotti, M., Lamy-Bergot, C., Huusko, J., Amon, P.: Content adaptive network aware joint optimization of wireless video transmission. IEEE Commun. Mag. **45**(1), 84–90 (2007)
8. Gallager, R.: Low-density parity-check codes. IRE Trans. Inf. Theory **8**(1), 21–28 (1962)
9. Planete-bcast, inria, ldpc codes download page. http://planete-bcast.inrialpes.fr/article.php3?id_article=16
10. Wang, Z., Bovik, A., Sheikh, H., Simoncelli, E.: Image quality assessment: from error visibility to structural similarity. IEEE Trans. Image Process. **13**(4), 600–612 (2004)

Author Index

Abdel Salam, A.   34
Apollonio, Pietrofrancesco   76

Bacco, Manlio   114
Birrane III, Edward J.   58
Bisio, Igor   123

Caini, Carlo   76
Caviglione, Luca   34, 114
Cello, Marco   89

Davoli, Franco   1
Del Re, Enrico   45

Fan, Yuanyuan   22
Fanfani, Alessio   45

Giusti, Marco   76
Gotta, Alberto   34, 114

Kato, Nei   94

Lacamera, Daniele   76
Lavagetto, Fabio   123
Li, Hui   22
Liang, Qingzhong   22
Liu, Chao   22

Luglio, Michele   34
Luzzati, Giulio   123

Marchese, Mario   12, 89, 123
Miura, Ryu   94
Morosi, Simone   45

Nishiyama, Hiroki   94

Patrone, Fabio   89

Ronga, Luca Simone   45
Roseti, Cesare   34

Sun, Ruijin   106

Takaishi, Daisuke   94

Wang, Guangjun   22
Wang, Ying   106

Yin, Chong   106

Zampognaro, F.   34
Zeng, Deze   22

Printed in the United States
By Bookmasters